Proud Past
Bright Future

MFA Incorporated's First 100 Years
CHUCK LAY

Copyright © 2013 by MFA Incorporated

All rights reserved, including the right to reproduce this work in any form whatsoever without permission in writing from the publisher, except for brief passages in connection with a review. For information, please write:

The Donning Company Publishers
184 Business Park Drive, Suite 206
Virginia Beach, VA 23462

Steve Mull, General Manager and Project Director
Barbara Buchanan, Office Manager
Anne Burns, Editor
Nathan Stufflebean, Graphic Designer
Kathy Adams, Imaging Artist
Cindy Smith, Project Research Coordinator
Tonya Washam, Research and Marketing Supervisor
Pamela Engelhard, Marketing Advisor

Library of Congress Cataloging-in-Publication Data
Lay, Chuck.
Proud past, bright future : MFA Incorporated's first 100 years / by Chuck Lay.
    pages cm
ISBN 978-1-57864-853-5
1. MFA Incorporated. 2. Agriculture, Cooperative--United States--History. 3. Agricultural industries--United States--History. I. Title.
HD1485.M58L39 2013
334'.68309778--dc23
                          2013027098

Printed in the United States of America at Walsworth Publishing Company

Table of Contents

Dedication 4
President's Letter 5
Preface 6
Acknowledgments 8
Introduction 9
1. William Hirth and
   the Beginning of MFA 12
2. Convergence—Politics,
   Federal Farm Board, Cowden,
   Farm Bureau and Paybacks 38
3. MFA in the 1930s 58
4. Passing the Torch 68
5. Fred Heinkel Takes Charge 82
6. Political Distractions 114
7. Transfer of Power 134
8. The Agricultural Depression 152
9. Modern Management 164
About the Author 184

This book is dedicated to the farmers and ranchers who had courage enough to change their destinies and to the people of MFA who give so much.

Dedication

So without premeditation, I leaned at him and the words sprang from my lips without consideration of courtesy: "But you are old, and you have made many crops," I said bitterly, "and does it please you to make another? Just like all the rest?"

He looked at me as though he thought me somewhat mad or as though I spoke Greek to him and him unlearned, and I waited for his reply. Then he slowly said, "Son, I made me many a crop. Some good and some poor to starvation. But I aim, under God's hand, to make me one more." He waited and took a sip from his glass, and then said: "And this one—it may be the best I ever made."

Robert Penn Warren
"World Enough and Time"

President's Letter

Thank you, Chuck Lay, for writing "Proud Past, Bright Future," a review of MFA's first 100 years of operation. It is an excellent read.

What isn't written down is lost to history. It is important that the stories of families, counties, institutions and corporations be preserved. There is much to celebrate from these stories, and much to learn. Given that MFA Incorporated is the oldest regional farm supply and grain marketing cooperative in the United States, we felt it paramount that the history of our business be perpetuated.

Historians will want a source to interpret the facts—to explain how things were and how they came to be. As MFA turns 100, we thought you might, too. That's how this book came to be.

There have been two other books written about the history of MFA. Ray Derr's 1953 book, "Missouri Farmers in Action," covers the cooperative's activities from 1914 to 1951.

Ray Young's 1995 book, "Cultivating Cooperation," reflects on MFA history from 1914 to 1979.

In the pages of this book, Lay has compiled the history of MFA which began March 10, 1914, when seven farmers met at Newcomers Schoolhouse to launch MFA, and followed that history to present day, 2013.

Lay is the director of communications at MFA Incorporated. His employment with MFA began in 1988 as editor of the Today's Farmer magazine, and he has shown a keen interest in the history of MFA since joining the company. It was only logical that he should take up the challenge of compiling documented history and add his original research to bridge the company's endeavors since 1979.

Chuck spent many hours at the Western Historical Manuscript Collection of the Missouri State Historical Society at the Ellis Library in Columbia, Mo.—the location housing the collected papers of William Hirth and Fred Heinkel. This collection provides an in-depth account of the first two MFA presidents (and great farm leaders). Their tenure spanned 65 years of MFA history. Much of Lay's research provides a unique perspective not previously reported.

In addition to reviewing, investigating and researching the archived files, Lay conducted extensive interviews with current and past employees of MFA, including Ray Young, Eric Thompson, Bud Frew and Don Copenhaver—four important past MFA leaders with a deep knowledge of the cooperative's inner workings and personalities.

MFA remained a very complex organization until 1983 when the business activities of the company came under one umbrella. As late as 1965, the MFA organization had 21 divisions and agencies. Also, the company had full or partial ownership of 15 additional companies and supported the efforts of 150 local MFA-affiliated cooperatives. Most of the MFA agencies had interlocking boards.

Through time, MFA has had strong and differing personalities in leadership roles. But it should not be forgotten that the organization was conceived as a business to provide economic benefit for farmers at a reasonable profit. In the 1950s the organization lost focus of its initial purpose and became consumed with legislative and political activities—at the expense of sound business direction and strategies.

It was not until the early 1980s that the company returned to the mission and vision of the original founders: providing goods and services to member owners while concentrating on sound business principles.

MFA has experienced internal political struggles, financial roller coaster rides and differing strategic directions that were counterproductive to good financial results. And, yet, we are here to tell about it.

Lay has done a masterful job of capturing the dynamics of this great company, bringing readers through our challenges and our success. The cooperative members and employees of MFA have a lot to be proud of.

MFA has a past, but also has an exciting present and a promising future.

Again, thanks, Chuck.

Bill Streeter, MFA President and Chief Executive Officer

Preface

As a writer, I've always been fascinated by historical context. The past as prologue. A glance backward to see forward. One of the many benefits of my job at MFA is that I have associated with tremendous leaders. In what seems an incredibly short span of time, I've met and come to know Ray Young, Eric Thompson, Bud Frew, Don Copenhaver, David Jobe, Janice Schuerman, Brian Griffith and Bill Streeter. These individuals will be followed by others history will kindly identify for us.

I don't want to be presumptuous, but I've seen the making of MFA history first hand. I continue to be impressed by the fierce intellect of these individuals. They make no pretense of intellectualism. Business is supposed to occur in the realm of practicality. In agriculture, we deal with farmers and the people who serve them. There's beauty in simplicity; genius in haiku. These individuals have amazing ability to focus on the abstracts of balance sheets and the power of objective data.

When I came to MFA in 1988, I unearthed scads of historical documents, researched in part by Ray Young in preparation for his book, "Cultivating Cooperation." Other documents languished in old files. I discussed MFA's history in-depth with him. For some reason, maybe he sensed my interest was genuine, he confided in me on a range of issues, told me events he'd witnessed and idiosyncrasies of people he'd known. Young sadly lamented that Fred Heinkel, MFA president from 1940 to 1979, had not known when to step down.

He thought the political turmoil and the associated confrontations were unseemly as well as too vivid. He would not detail the subject in his forthcoming book, he told me. He was too close, the subject still too sore. He would leave the matter for the next writer to deal with, hoping for historical objectivity. "Maybe, you," he said and winked.

I had been hired as editor of Today's Farmer in 1988 after interviewing with Bill Streeter, who, at the time, was vice president of sales. He patiently explained the purpose of Today's Farmer was to provide helpful information to farmers. He understood the history of the magazine (which predates the cooperative by six years) and its significance. "Your job is not sales," he said. "Your job is providing factual information that will help farmers make responsible decisions. When farmers make money, MFA makes money." It remains the objective of Today's Farmer.

The first day on the job, I met then President Bud Frew in his office. He told me I would be given a free hand unless I had an agenda. He told me his door was always open but in typical blunt fashion he ended with: "But don't abuse the privilege." Don't misunderstand. Bud Frew was a practical man, but he balanced those forthright manners. He was kind and an excellent mentor. He took me to meetings my position did not merit. He wanted me to understand the significance of these meetings, he said, what the cooperative stood for and his motivation in leading MFA. Bud Frew was complex. He could be warm and engaging, but he was always intensely focused on doing the right thing for the right reason.

When I first met former MFA President Eric Thompson, he was dying of cancer but eager to explain the significance of his tenure at MFA. "Chuck, I know people," he told me. "I tend to size them up early. You look to be a trustworthy s.o.b. You write it the way it happened." He opened his personal files to me—gave me complete possession of them because he understood their significance to MFA's history. And he sat through hours of interviews despite battling his fatal illness.

I interacted with Don Copenhaver and Bill Streeter from day one. Both are honorable men who in their own ways focused intently on keeping MFA in the business of serving farmers. Fred Heinkel died in 1990. He'd been in a nursing home the last year of his life. I still kick myself for missing the opportunity to meet him, interview him and learn his story in his words.

MFA is a story of hundreds if not thousands of individuals who have contributed in varying degrees. This book details only the activities of a few who had leadership roles. Records of individual performance are slim. Like all of human history, majorities have disappeared while the records of a few individuals have been retained. Ray Young did a yeoman's job of listing individual MFA managers, but still, as to individual farmers, he barely scratched the surface. Pyramids are named for kings, but the architects, engineers, craftsmen and laborers who contributed are lost to history. I encourage young people to dig into the histories of farm clubs and local exchanges in their areas. They'll find their own stories and maybe the stories of their ancestors.

And, of course, the real contributions come from those farmers and ranchers who saw opportunity and conquered traditional practices to build greatness. Farm clubs came first, and individual farm club members reached to others in their communities and built exchanges. In like fashion, they also reached outside their communities to build larger businesses still: creameries, livestock shipping associations, produce exchanges and the like.

Outside of internal MFA files, I have relied on two books covering MFA's history. They are: "Cultivating Cooperation" by Raymond A. Young and "Missouri Farmers in Action" by Ray Derr. As a result of the detail in listing individuals, MFA companies and dates by Young and Derr, I have not tried to recreate their efforts. This book is more narrative detailing highlights.

Two other important books in the research process are: "History of the Missouri Farm Bureau Federation" by Vera Busiek Schuttler and "Beyond the Fence Rows, A History of Farmland Industries, Inc." by Gilbert C. Fite. Several academic papers have been published, the best of which is "Bill Hirth and MFA: The Early Years" by V. James Rhodes. Two others I relied on are: "William A. Hirth: Middle Western Agrarian," by Theodore Saloutos and "The Politics of Desperation: William A. Hirth and the Presidential Election of 1932" by Richard O. Davies.

The importance of the Western Historical Manuscript Collection of the State Historical Society at the University of Missouri cannot be overstated. Many of the quoted and printed documents are from those archives.

I have left internal spelling and punctuation as is, even in speeches, since those speeches exist in written form. The periods after the M, the F and the A have, in quoted materials, been retained. Until 1981, the letters were the abbreviation for the Missouri Farmers Association. In 1988, MFA's articles of incorporation were officially modified to change the name of the cooperative to MFA Incorporated, which better reflects the multi-state presence of the modern cooperative.

Above all, let me end with this: Integrity is a theme running through MFA's modern history. Having interviewed past board presidents, past vice presidents, and current and past presidents, I can say with certainty the word "integrity" surfaced often. Especially in today's world, that moral compass shines all the more brightly. Those I have met here and reported to have added a great deal of luster to that shine.

Chuck Lay

Acknowledgments

I owe a heartfelt debt to the people who helped make this project possible. Bill Streeter, MFA president and CEO, was always enthused and offered excellent advice. Janice Schuerman is senior vice president of corporate services. She constantly ran interference for me and was understanding of the book's deadline constraints. MFA's communications department put up with my insistent requests for help. Nikki Larimore was helpful and quick to respond on last-minute details; Craig Weiland, art director extraordinaire, designed the dust jacket; Steve Fairchild stepped in to pick up projects I had no time for; Austin Black scanned hundreds of photographs and took many of the studio shots; and James Fashing, master photojournalist and technological whiz, designed MFA's image archive system and spent countless hours sorting, retrieving and manipulating the images you'll see in this book. He also took most of the photographs from 1994 to date. Erin Teeple kept grammatical watch, and Larna Lavelle was a wealth of information on the corporate board.

Introduction
The Business of Agriculture

MFA stands as testament to the ingenuity, hard work and entrepreneurial spirit of Midwest farmers and ranchers. Founded by seven farmers in 1914, MFA quickly grew into a Midwest economic powerhouse. As a farm supply and marketing cooperative in Missouri and adjacent states, MFA serves more than 45,000 farmers and ranchers.

It does so through a unique arrangement of MFA Agri Services Centers, locally owned affiliates, privately owned enterprises and agribusiness partnerships. MFA's retail presence provides member-owners and customers with products and services essential for crop and livestock production. MFA Incorporated's wholesale divisions provide a full range of agricultural inputs to customers within and outside of the cooperative.

In 2012, MFA ranked ninth nationally in retail, seventh in grain, eighth in storefronts, eighth in fertilizer sales, 10th in crop protection sales, eighth in custom application, eighth in seed and fifth in precision ag. For 100 years, MFA has grown by meeting the needs of farmers and ranchers as their operations evolve. Through its retail network, MFA delivers the best in agricultural inputs and services from top field-crop genetics and precision agriculture to grain marketing, farm supply, feed, animal health, plant foods, marketing services and credit solutions.

MFA has formed joint ventures with existing companies to build on geographic and other benefits such arrangements offer. Partnerships with grain and plant-foods facilities give MFA member-owners more access to retail opportunities. MFA has also formed partnerships with wholesale fertilizer supply

HEPLER AGCHOICE
Quick stats

- 3 semi-tenders
- 1 ten-wheeler
- 4 semi-tractors
- 46 anhydrous tanks
- 8 spreaders
- 1 floater truck
- 1 four-bin floater truck
- 1 sprayer

Recent improvements

- New from the ground-up is Hepler's AGChoice retail facility and warehouse. It opened for business in early 2012.

AGChoice at Hepler, Kan., opened for business in early 2012. The AGChoice retail facility is an outlet for plant foods, crop protection products, seed, farm supply and feed. AGChoice at Hepler moves more than 10,000 tons of dry plant foods, $300,000 of farm supplies and 4,000 tons of feed to area farmers and ranchers. The facility is one of multiple AGChoice locations owned by MFA Enterprises and located in Kansas, Oklahoma, Missouri and Iowa.

businesses and seed production facilities with the continuing goal of increasing buying efficiencies for customers.

In order to be a leading, full-service agricultural supplier, MFA reaches the length of the supply chain. MFA's annual plant food sales approach or exceed one million tons each year. Strategic river terminals and other bulk facilities in the sales territory give MFA the capacity to deliver bulk quantities of plant foods to MFA's retail outlets and dealer customers.

MFA still lives by its founding principles: strong corporate support is vital to each local MFA Agri Services Center or related business; timely communication with member-owners is essential; and service is all-important. Above all, MFA is successful as a company only when its individual member-owners achieve success in their operations as well.

The history in this book is an attempt to chronicle the spirit of the cooperative from its beginnings to the vibrant company of today.

LEBANON MFA AGRI SERVICES

Quick stats
- 47 fertilizer carts
- 5 multi-compartment feed trucks
- 5 spreader trucks
- 4 tender trucks
- 3 No-till drill

Farmers Produce Exchange #139 is a locally owned cooperative that has been affiliated with MFA since July 24, 1920. Lebanon's early affiliation included being one of the first sites of an MFA Oil Company bulk plant in 1930 as well as the site of an early creamery. With 47 fertilizer carts and multiple feed, spreader and tender trucks, the cooperative is geared to serve the farmers and ranchers who own it. Livestock and forage are major parts of the business, which includes retail branch locations at Conway and Grove Springs as well retail facilities at Richland.

BOONVILLE MFA AGRI SERVICES
Quick stats

- 30 fertilizer carts
- 55 anhydrous nurse tanks
- 12 tool bars
- 1 precision fertilizer spreader
- 1 post fertilizer spreader
- 1 fertilizer and lime spreader
- 5 post sprayers
- 1 floater sprayer
- 6 dry tenders
- 1 truck chassis seed tender
- 6 road tractors
- 1 hopper bottom

Recent improvements

- Bulk fertilizer facility
- Bulk seed storage
- Bulk crop protection storage

With 30 fertilizer carts, 55 anhydrous nurse tanks, 12 tool bars and multiple types of spreaders and sprayers, the Boonville MFA Agri Services Center is geared up for crop production. Boonville also features bulk facilities for plant foods, seed and crop protection products—all structured to serve farmers and ranchers. The Boonville MFA has seen steady growth in sales and continues to upgrade facilities accordingly. The Boonville MFA has been part of the MFA system since the 1920s.

William Hirth was the driving force behind the creation of MFA.

★★ Chapter 1 ★★
William Hirth and the Beginning of MFA

A human soul on fire

William Hirth burned brightly. A first-class orator, a genius of organization, a national political leader, Hirth bent the world to his will. It was a capacity that brought him acolytes as well as critics, both in his time and since. Alternately, he was vilified as domineering, utterly without humility and bellicose by his critics, especially those in academia.

Yet to his supporters, Hirth, the farm boy who rose from obscurity on a Rush Hill farm in central Missouri to become a national farm leader who dined with presidents and senators, was a dynamo of leadership, a man obsessed with changing the lives of farmers for the better. Hirth's vision? The creation of the farm organization which became MFA—a battle he refused to lose.

William Hirth created the Missouri Farmers Association, one of the 20th century's most successful and dynamic cooperatives and one still thriving in the 21st. Seven farmers from the Newcomer Schoolhouse Farm Club near Brunswick are rightly praised as the genesis of the organization. But make no mistake, they followed the template Hirth forged. Hirth wrought that template (even to the extent of conceiving the famous shield logo himself) through passion, persuasion, skill and no small amount of luck.

Born March 23, 1875, Hirth was raised on the family farm in Audrain County, Mo. His experience of the soul-dampening drudgery of 1800s farm life was acquired firsthand. As a young man, Hirth saved enough money to finance a year's tuition at McGee College in Macon. A year later, he transferred to Central College in Fayette. His money ran out in two years. Nearly penniless, he left college and began selling insurance. He prospered. By 1900, he and his new wife, Lillian Vincent, moved to Columbia where he read law and was admitted to the bar.

In 1906 he purchased the Columbia Statesman newspaper. But agriculture continued to hold his interest. By 1908 he realized his dream of developing a statewide agricultural publication when he purchased The Missouri Farmer and Breeder, now Today's Farmer. The first issue was published October 15, 1908. In February 1912, he shortened the name to The Missouri Farmer.

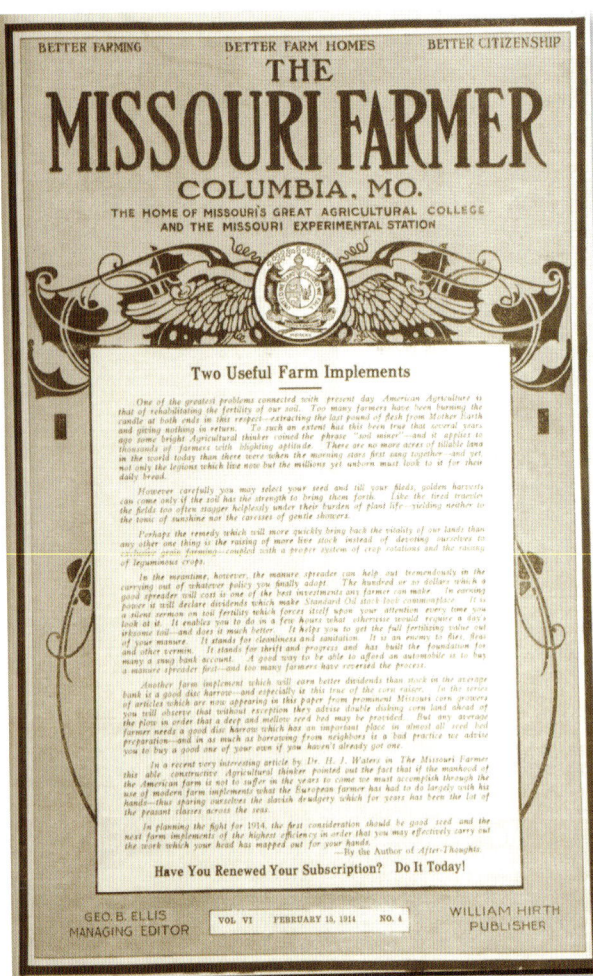

By 1914, The Missouri Farmer, widely popular in state agricultural circles, was the official publication for the Missouri Farm Management Association, the Missouri Cattle Feeders Association, the Missouri Corn Growers Association, the Missouri Draft Horse Breeders Association, the Missouri Dairy Association and the Missouri Sheep Breeders Association. Hirth owned the tabloid magazine until his death in 1940.

At the time, the U.S. Census Bureau, newly ensconced in the just-created U.S. Department of Commerce and Labor, counted 92 million citizens in the 46 states. For historical perspective, consider: The first ship had yet to steam through the newly built Panama Canal. When Hirth named his magazine The Missouri Farmer, the HMS Titanic was two months shy of finding the iceberg. Orville and Wilbur Wright had defied gravity just nine years earlier. Henry Ford was still a year away from mass production of the Model T. Horse and mule teams powered agriculture. One quarter to one third of farm acreage was devoted to oats and forage crops for those teams.

◀ By calling for the formation of farm clubs, William Hirth used his magazine to launch the cooperative that became MFA.

▼ At the dawn of the 20th century, horse and mule teams powered agriculture. Hirth preached cooperation, farm clubs and exchanges to provide economic power to isolated farmers.

The Missouri Farmer provided Hirth a platform to evangelize. Print was king. The first commercial radio broadcast was eight years in the future. Telephones were nonexistent. Electricity was enjoyed only by select urban elite.

The Missouri Farmer hit a rural nerve and drew instant response. Hirth's passion was organization, which he saw as key to improving farm life. Hirth preached organized cooperation repeatedly in its pages. Cooperation in terms of a business structure had been introduced early in America by Benjamin Franklin. History proved the principle sound. Aaron Bachtel, a farmer and stockman just north of Brunswick, was a subscriber who took Hirth's ideas seriously.

"In the February 1914 issue," Bachtel said, "there was a short article wherein it stated that farmers should organize into school district Farm Clubs and how they could be benefitted by such organizations. It looked so simple and at the same time so far reaching, that it appealed to me very forcibly."

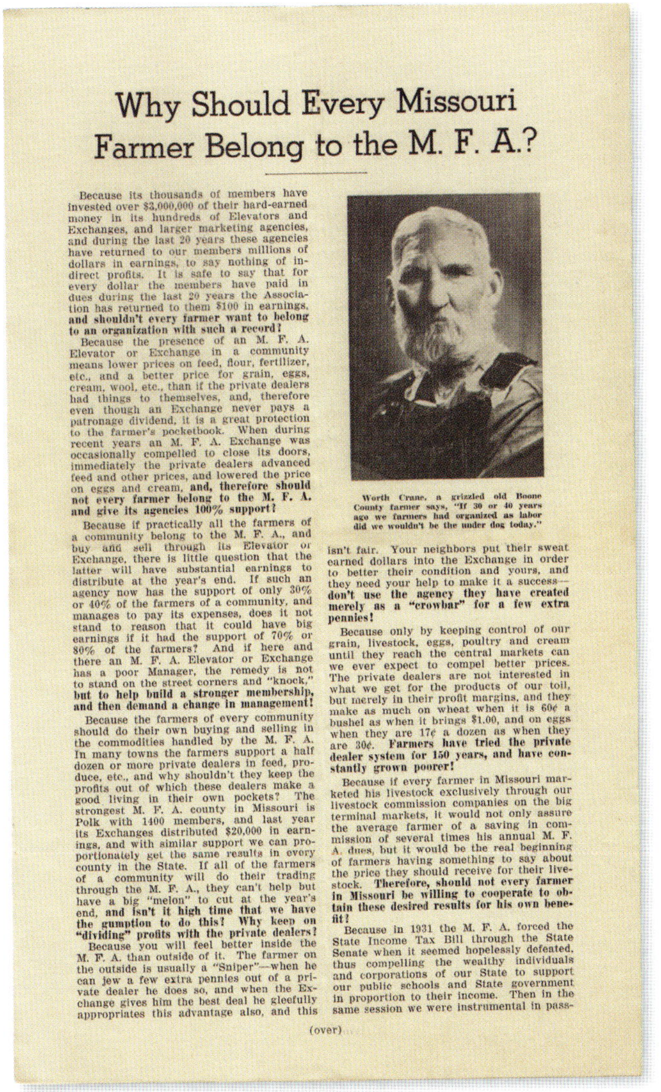

Why Should Every Missouri Farmer Belong to the M. F. A.?

Because its thousands of members have invested over $3,000,000 of their hard-earned money in its hundreds of Elevators and Exchanges, and larger marketing agencies, and during the last 20 years these agencies have returned to our members millions of dollars in earnings, to say nothing of indirect profits. It is safe to say that for every dollar the members have paid in dues during the last 20 years the Association has returned to them $100 in earnings, **and shouldn't every farmer want to belong to an organization with such a record?**

Because the presence of an M. F. A. Elevator or Exchange in a community means lower prices on feed, flour, fertilizer, etc., and a better price for grain, eggs, cream, wool, etc., than if the private dealers had things to themselves, and, therefore even though an Exchange never pays a patronage dividend, it is a great protection to the farmer's pocketbook. When during recent years an M. F. A. Exchange was occasionally compelled to close its doors, immediately the private dealers advanced feed and other prices, and lowered the price on eggs and cream, **and, therefore should not every farmer belong to the M. F. A. and give its agencies 100% support?**

Because if practically all the farmers of a community belong to the M. F. A., and buy and sell through its Elevator or Exchange, there is little question that the latter will have substantial earnings to distribute at the year's end. If such an agency now has the support of only 30% or 40% of the farmers of a community, and manages to pay its expenses, does it not stand to reason that it could have big earnings if it had the support of 70% or 80% of the farmers? And if here and there an M. F. A. Elevator or Exchange has a poor Manager, the remedy is not to stand on the street corners and "knock," **but to help build a stronger membership, and then demand a change in management!**

Because the farmers of every community should do their own buying and selling in the commodities handled by the M. F. A. In many towns the farmers support a half dozen or more private dealers in feed, produce, etc., and why shouldn't they keep the profits out of which these dealers make a good living in their own pockets? The strongest M. F. A. county in Missouri is Polk with 1400 members, and last year its Exchanges distributed $20,000 in earnings, and with similar support we can proportionately get the same results in every county in the State. If all of the farmers of a community will do their trading through the M. F. A., they can't help but have a big "melon" to cut at the year's end, **and isn't it high time that we have the gumption to do this? Why keep on "dividing" profits with the private dealers?**

Because you will feel better inside the M. F. A. than outside of it. The farmer on the outside is usually a "Sniper"—when he can jew a few extra pennies out of a private dealer he does so, and when the Exchange gives him the best deal he gleefully appropriates this advantage also, and this isn't fair. Your neighbors put their sweat earned dollars into the Exchange in order to better their condition and yours, and they need your help to make it a success—**don't use the agency they have created merely as a "crowbar" for a few extra pennies!**

Worth Crane, a grizzled old Boone County farmer says, "If 30 or 40 years ago we farmers had organized as labor did we wouldn't be the under dog today."

Because only by keeping control of our grain, livestock, eggs, poultry and cream until they reach the central markets can we ever expect to compel better prices. The private dealers are not interested in what we get for the products of our toil, but merely in their profit margins, and they make as much on wheat when it is 60¢ a bushel as when it brings $1.00, and on eggs when they are 17¢ a dozen as when they are 30¢. **Farmers have tried the private dealer system for 150 years, and have constantly grown poorer!**

Because if every farmer in Missouri marketed his livestock exclusively through our livestock commission companies on the big terminal markets, it would not only assure the average farmer of a saving in commission of several times his annual M. F. A. dues, but it would be the real beginning of farmers having something to say about the price they should receive for their livestock. **Therefore, should not every farmer in Missouri be willing to cooperate to obtain these desired results for his own benefit?**

Because in 1931 the M. F. A. forced the State Income Tax Bill through the State Senate when it seemed hopelessly defeated, thus compelling the wealthy individuals and corporations of our State to support our public schools and State government in proportion to their income. Then in the same session we were instrumental in pass-

(over)

▲ Hirth used his printing press to spread the word on cooperation. Farmers and ranchers used Hirth's medium to encourage neighbors and form farm clubs. (Western Historical Manuscript Collection)

► Aaron Bachtel was a farmer and stockman in central Missouri. He and six neighbors formed the Newcomer Schoolhouse Farm Club in 1914, pooled an order of twine and sent that order to William Hirth.

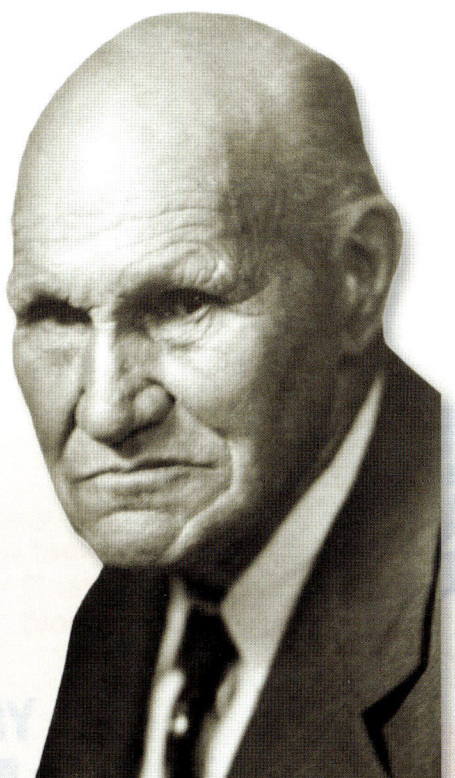

"It looked so simple and at the same time so far reaching, that it appealed to me very forcibly."

—Aaron Bachtel

The Newcomer Schoolhouse still stands today outside of Brunswick, Mo. A reproduction of it is located in what is now Shelter Gardens in Columbia, Mo.

Bachtel wrote Hirth, asking for a bundle of the papers to distribute to his neighbors. He asked those same neighbors to meet at the Newcomer Schoolhouse the following Tuesday night.

A prominent farmer, Bachtel lent his name and reputation to the cause. He asked his neighbors to unite in the purchase of farm supplies and inputs. By seizing the initiative, Bachtel overcame local doubt and skepticism to convince others that Hirth's idea to organize farmers around a cooperative concept was sound. The group pooled an order for 1,150 pounds of penitentiary binder twine and sent it to Hirth in Columbia. The Newcomer Schoolhouse Farm Club volume buy netted the men $400 in savings.

That Newcomer Schoolhouse Farm Club order, the first Hirth received, counted as the creation of the Missouri Farmers Association. It marked the beginning of what grew into a cooperative devoted to pooling orders and distributing accumulated savings. The farm club idea swept the countryside, according to historians, like a prairie fire. Subsequent to that first order, Hirth contracted with the West Virginia Coal Company at a set price for coal. Bachtel and his neighbors were on board.

Soon after the order, coal prices jumped, but the fledgling MFA had a contract at the cheaper price. Bachtel received several carloads and proceeded to instruct area farmers in the value of the farm clubs. As Bachtel later related: "I told all the farmers, regardless of whether they were members or not, to come in and get coal just the same, with one provision, that when they got their coal, if they thought the Farm Club was really worth while they were to pay their dues and become members, which they did, all but one man." Bachtel would pause in the telling and add: "And I've never liked him since."

The members of the Newcomer Schoolhouse Farm Club at that first meeting were: Aaron Bachtel, T.E. Penick, brothers W.J. and George Heisel, Earl Smutz, John Kohl and W.L. Armstrong. Ten more members quickly flocked to the club. Throughout the state, the scene was repeated, and repeated, and repeated again.

For perspective on the prairie-fire metaphor, consider: Within five years, approximately 1,000 MFA farm clubs were created representing every corner of the state. Just over 10 years later, Hirth could announce the existence of hundreds of exchanges, elevators, central produce plants, creameries, livestock shipping associations, livestock commission companies, a grain commission company, a purchasing department and produce sales agencies. All told these fledgling creations dedicating themselves to MFA were generating gross sales of more than $100 million.

D. C. McCLUNG, WARDEN.   OLIVER BASSMAN, CHIEF CLERK.   PORTER GILVIN, DEPUTY WARDEN.

MISSOURI STATE PENITENTIARY
WARDEN'S OFFICE.
JEFFERSON CITY, MO.

June 8, 1916.

Mr. Aaron Bachtel,
Brunswick, Mo.

Dear Sir:

Replying to your letter of the 7th inst., will state that we have added another 50 pounds of twine to your order, making a total of 1050 pounds, to be shipped when you give us shipping directions. You may either send the money for this or we will ship same sight draft bill of lading attached.

Yours truly,

D. C. McClung
WARDEN.

DCM/M.

This original document from the MFA files is an early order from Aaron Bachtel and the Newcomer Schoolhouse Farm Club. The first order in 1914 was for 1,150 pounds of binder twine.

Proud Past, Bright Future: MFA Incorporated's First 100 Years   17

An early, but undated, MFA board of directors 1) J.M. Jones, Everton; 2) R.M. Wood, Lamar; 3) A.J. Crawford, Atlanta; 4) W.A. Kearnes, Granger; 5) George Kelley, Tipton; 6) F.M. Carrington, Safe; 7) Fred Miller, Holden; 8) C.M. Cope, Crane; 9) Aaron Bachtel, Brunswick; 10) Chris Ohlendorf, Boonville; 11) T.M. Chapman, Ozark; 12) W.S. Miller, St. James; 13) J.L. Pritchard, Knob Noster; 14) J. Wiley, Atkins; 15) T.H. DeWitt, Green City; 16) Herman Hetlage, Wright City; 17) C.M. Stiles, Bolivar; 18) unidentified; 19) Howard Cowden, Columbia; 20) unidentified.

Bachtel would serve as president of MFA's first farm club and become the first farmer to sign MFA's unique, but doomed to fail, producer contract. Bachtel would log 20 years of distinguished service on MFA's corporate board of directors.

Through the pages of The Missouri Farmer, Hirth chronicled farmer efforts and savings of farm clubs, urging continued development of more clubs where farmers could organize and socialize in much the same way town businesspeople joined business, political and benevolent organizations. Self-interest, said Hirth, should be a natural draw for historically independent farmers.

Yet organizing farm families was a steep hill as Hirth reflected in a radio speech in 1931: "And there are reasons why farmers are hard to organize. In the first place, they live somewhat isolated lives. The laborer or business man in town is in almost hourly touch with his fellows, while often except to talk to a neighbor across the fence, or when he passes him on a public road, the farmer lives in the bosom of his family and in his fields and feedlots for a week or more at a time."

Hirth, nevertheless, was wildly successful in convincing farmers of the benefits of cooperation, due in main to the force of his personality and his frenetic, but strategic, schedule. In May of 1920, Hirth was the principal speaker at the Barton County farm clubs' meeting in Lamar. The Lamar Democrat covered his speech and concluded "Hirth made a profound impression." In 1921, Hirth appeared in Warsaw in the afternoon, Cole Camp that evening, Versailles the following morning. His schedule took him to Stockton, Browning, Treloar, Flint Hill, Kirksville, St. James as well as multiple towns in Franklin, Macon, Shelby and Lewis counties in a matter of weeks.

Local buildings could not contain the crowds. Newspapers described small-town traffic jams, farm families waiting hours for entrance and a rural audience enthralled.

"And there are reasons why farmers are hard to organize. In the first place, they live somewhat isolated lives. The laborer or business man in town is in almost hourly touch with his fellows, while often except to talk to a neighbor across the fence, or when he passes him on a public road, the farmer lives in the bosom of his family and in his fields and feedlots for a week or more at a time."

—William Hirth, 1931 radio speech

MFA Stores

Adrian

Aldrich

Argyle

Aurora

Barnett

▲ The Women's Progressive Farmers Association was the women's affiliate of MFA. First formed in 1920, WPFA elected officers, published news articles in The Missouri Farmer and served meals at MFA meetings. WPFA held meetings in conjunction with MFA. WPFA published cookbooks that are still in demand and had food canning contests.

◄ Convincing farmers and ranchers to improve their lot meant breaking from tradition, as this poem, printed in the women's affiliate organization WPFA, cleverly points out. (Western Historical Manuscript Collection)

The Slogan

"What was good enough for Father
    Must be good enough for me,"
Said the dinosaurus paddling
    In the ancient shipless sea;
Which is why he's in Museums,
    While the more progressive cow
Has electric light and oil cake
    In a model stable now.

So whatever in the future
    You may set yourself to do,
Making airships, building houses,
    Hunting microbes in the blue,
Rearing babies, dealing justice—
    Let your slogan ever be
"What was good enough for Father
    Is not good enough for me."
            —Exchange.

Farm clubs structured meetings, printed songbooks, formed a women's auxiliary organization (Women's Progressive Farmers Association or WPFA in 1921) and scheduled debates. As Ray Young's book describes, hundreds of clubs held debates every two weeks, organized fishing trips, held parades several miles long and hosted barbecues. The spirit of a camp revival was in the air.

At those meetings Hirth encouraged farmers to spread the gospel, noting that although many farmers were uncomfortable in the speaker's role, it was a mission worth the effort. True oratory, he intoned, was not faultless diction or learned phrases. True oratory "is a human soul on fire with a great cause."

"I'll buy it from the people who made you lower yours."

Counties could and did contain multiple farm clubs that claimed membership in the Missouri Farmers Association. Over time, those clubs in turn organized (or federated) into county units. And one by one those units coalesced into a statewide organization, yet were owned lock, stock and barrel by the local members who built facilities, hired and fired managers and elected local boards.

True oratory 'is a human soul on fire with a great cause.'

—**William Hirth**

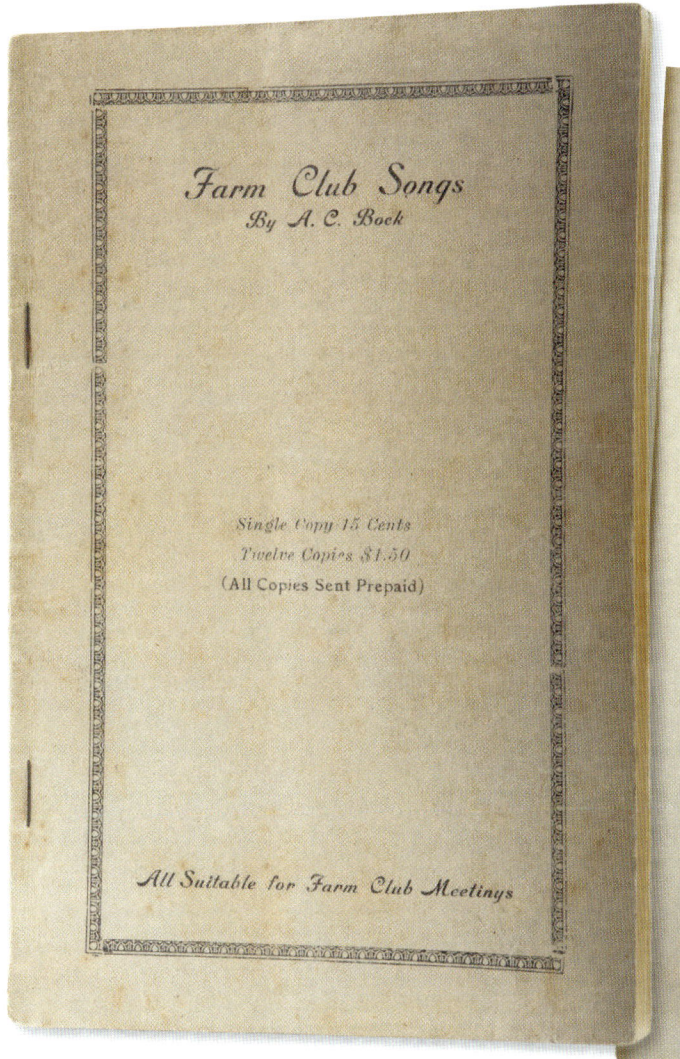

▲ MFA members wrote and submitted dozens of MFA-themed songs and songs of farm life. Farm-club meetings and MFA conventions devoted program time to rousing songs.

◀ Hirth saw organization as key to giving farmers a voice in the marketplace. He frequently noted that U.S. businesses presented a united front through the Chamber of Commerce and labor was organized and represented by unions. Farmers needed the power provided by cooperatives if they were to command economic power.

By 1920, more than 200 MFA exchanges had been organized. Those exchanges, in turn, accounted for 100 grain elevators, 150 livestock shipping associations and poultry and egg processing plants around the state.

By 1919 (just five short years), MFA was handling one-fourth of the entire state's needs for binder twine, all of which had been ordered through more than 1,000 farm clubs. Growth was exponential. By 1920, more than 200 exchanges accounted for 100 grain elevators, 150 livestock shipping associations as well as poultry and egg processing plants at Springfield, Medill and St. Joseph.

Total U.S. crop acreage stood at 370 million acres, farmed by 6.5 million farmers. Cotton sold for 12 cents a pound. In 1921, to market their livestock and produce, MFA supporters founded Farmers Livestock Commission in East St. Louis and, in Springfield, Producers Produce Company (poultry, eggs and produce). Producers Creamery Company (milk products) began in Springfield in 1927. Each individually became the nation's largest. By 1929, MFA, responding to farmer demand for kerosene, gasoline, tires and oil, began MFA Oil Company.

Historians attribute MFA's phenomenal success to Hirth's focus on providing immediate financial savings, a practice not employed by competing groups of the day nor of years prior. His papers on file at the Western Historical Manuscript Collection at the University of Missouri contain hundreds of pages of correspondence with suppliers vying for MFA's business.

Each notes the buying power of the new farm-supply powerhouse and offers large volume discounts, which Hirth immediately turned into farmer savings by bringing down prices significantly, putting farmers in charge of their destiny.

Included in the Hirth files at MFA Incorporated is an anecdote of a merchant telling his former farmer/customer that he'd happily match MFA's lower price. The farmer's response? "No thanks. I'll buy it from the people who made you lower yours."

Immediate accomplishments

Demonstrating the effectiveness of the power of centralized buying, offers poured in from flour mills, Ford and Dodge motor companies, tire manufacturers, seed merchants and new animal health businesses like Anchor Serum Company of St. Joseph.

Hirth, ever the skeptical farmer/businessman, immediately wrote Dr. J.R. Mohler, chief of the Bureau Animal Industry, U.S. Department of Agriculture in Washington, and asked for his opinion "as to the standing of the Anchor Serum Company of South St. Joseph."

And Hirth, being Hirth, was direct, if not confrontational with recalcitrant vendors, writing June 19, 1929, to the Morton Salt Company:

> The Missouri Farmers' Association operates 377 Grain Elevators and Produce Exchanges in this State, and is therefore the most powerful distributive machine of its kind in the Corn Belt. For a number of years we have had Contracts with leading flour, feed, fertilizer and twine manufacturers which give us inside terms, but whenever we have attempted to make such a Contract with a good Salt company we have been met with an attitude of mystery and reticence, and some of these companies have admitted to us confidentially that they are afraid of the retaliation of the Morton Salt company should they enter into such an arrangement.
>
> The time has come when we are determined to unearth the actual facts in these premises, and if necessary we will ask the Federal Trade Commission to inaugurate a thorough investigation. However, before we do this I have concluded to put the matter up to you in the hope that we may be able to reach a friendly understanding on this score with your Company or with some other responsible company. An organization of the size of the Missouri Farmers' Association is entitled to consideration in proportion to its distributive power and we shall insist upon such recognition.

Morton Salt hastened to assure Hirth that the rumors were unfounded. Salt contracts were forthcoming.

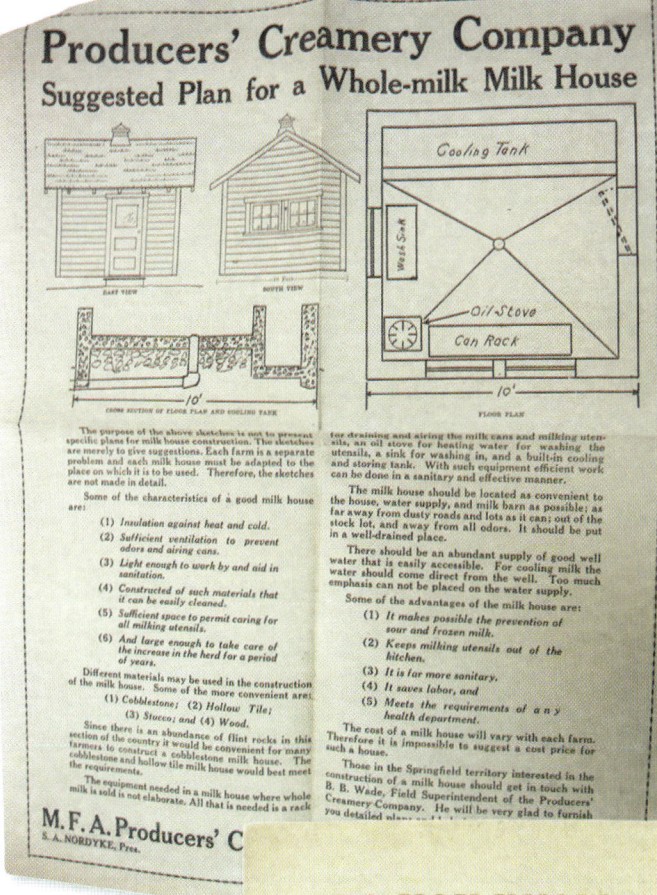

▲ To supply fresh milk and cream to the MFA creameries sprouting up statewide, MFA Producers Creamery led efforts to improve production facilities on the farm. Many of the native-stone buildings still dot the countryside around the Ozarks. (Western Historical Manuscript Collection)

▶ Hirth printed and circulated hundreds of publications designed for farmer education and membership recruitment. (Western Historical Manuscript Collection)

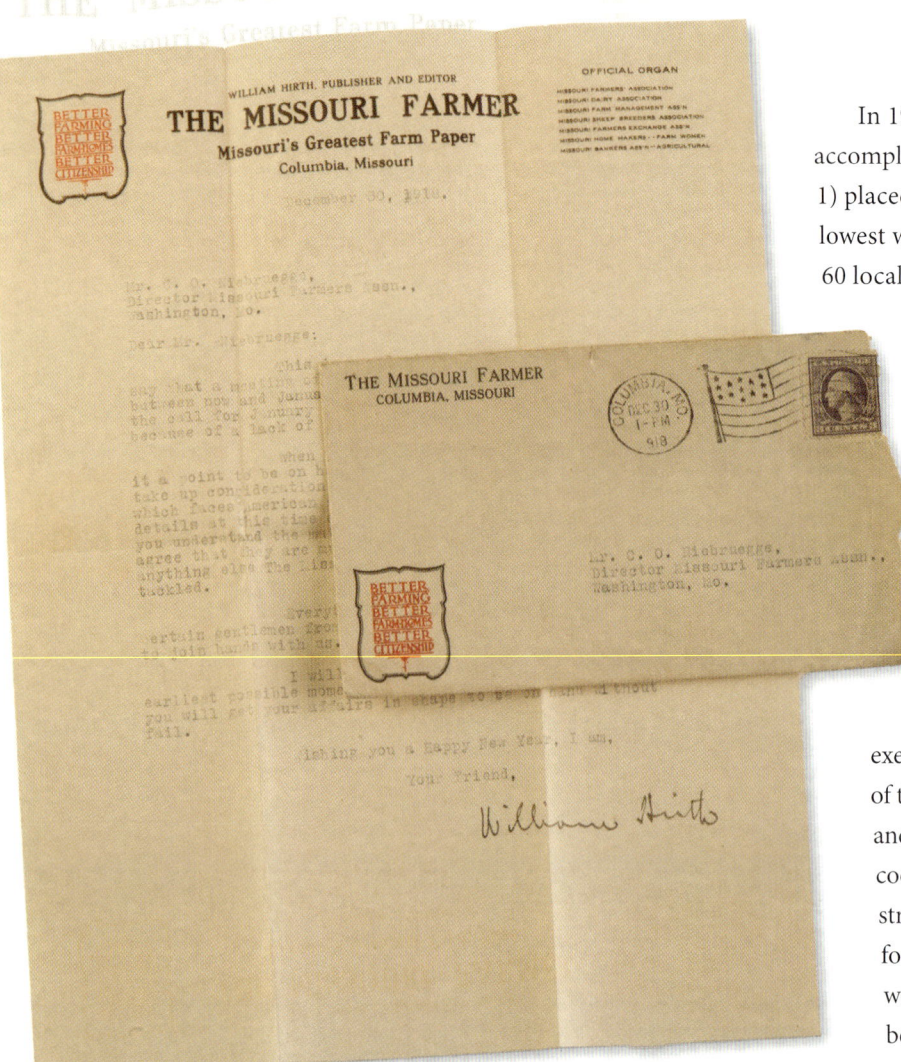

In 1919, Hirth listed the farm club accomplishments to date. Farm clubs, he wrote, have: 1) placed members in position to buy farm needs at lowest wholesale price; 2) financed and taken over 60 local elevators with plans for 150 before 1920; 3) been first to introduce a Live Stock Shipping Association in Missouri; 4) financed a flour mill costing $300,000; 5) secured regulation changes regarding the handling of meat under the Food Administration; 6) met with J. Ogden Armour as the first organized effort to work out a solution to packing interest; and 7) organized farmers to influence legislation.

By way of explanation of those accomplishments, Hirth, with the executive committee and the full board of directors of the nascent MFA, had lobbied newspaper editors and lawmakers for passage of an agricultural cooperative law allowing patronage refunds and streamlining cooperative practices. They also fought for and won government standardization of weights and measures for business scales that had been a bit haphazard before.

▲ Hirth kept up a steady flow of information to subscribers and members.

▼ This egg scale has different settings for small, medium and large.

24   WILLIAM HIRTH AND THE BEGINNING OF MFA

MFA sponsored a legislative dinner in Jefferson City to explain farmers' intentions in building self-help cooperatives. MFA's corporate board sponsored the law and saw to its passage by the newly lobbied legislature in 1919.

Just a year before, Hirth, never one to think small, took a group of official MFA representatives to Washington, D.C., to meet with the Agricultural Committee of the Senate to explain MFA's position on a federal cattle and hog feeder program. In each case, Hirth was upfront on intentions. MFA would be known for putting cards on the table frankly, in full view. No subterfuge for the new MFA.

Emboldened by success, Hirth led the group to Chicago to meet with J. Ogden Armour, titan of the meat-packing industry. Armour met with the group to discuss cattle and hog markets but made no promises. Hirth, however, made a friend with whom he corresponded in subsequent years. Despite no concrete proposals, Hirth still considered the trip a success. The farmers were flexing newly acquired muscle.

An MBA's nightmare

Looking at MFA's early development and subsequent structure, historians and perplexed academics declared MFA to be neither fish nor fowl. The original structure, lasting until mid-century, was baffling even to MFA's management. MFA was an organization. It was an association. It was almost a federated cooperative. It was almost a centralized cooperative. To varying degrees, the early Missouri Farmers Association was an unwieldy combination of all. But first and foremost, MFA represented the farmers of the state. For, after all, farmers and their farm clubs, not the various cooperatives, dictated policy.

Ray Derr, who wrote the first history of MFA in the 1950s, found the organization so complex even its leaders were unable to accurately describe the cooperative's structure. What's more, Derr noted, those same leaders "are sometimes at a loss to know the exact number of M.F.A. Exchanges, elevators, mills, creameries, and other units existing at a given time."

Why? Because Hirth's original vision did not include a corporate structure, but should have as subsequent events proved. Hirth's vision was to convince farmers to build an economic powerhouse. He was not drawing cooperative blueprints. For good or bad, he focused laser-like on farmers helping themselves. With Hirth behind the scenes, urging them on, conducting business prospectuses and studies, negotiating contracts, dealing with lenders and vendors, farmers were organizing first farm clubs, then exchanges, then structures to collect, market and distribute their products.

For Derr, MFA's history differed from other cooperatives in that, "it was to be specifically an economic agency…The objective was to secure a fair profit for the fruit of the farmer's sweat and toil—a phrase to be heard again and again during the early years of the Association—and all other activities were to be subservient to this ideal."

That initial web-like structure, created by a groundswell, was an MBA's nightmare. Yet it was effective, unbelievably so to the tune of hundreds of millions of dollars. By the 1930s, MFA was one of, if not the largest cooperative businesses in the nation. But not without conflict. Not without turmoil. Not without portent of future management trouble.

MFA Stores

Beaufort

Berger

Blackwater

Boonville

Branson

An avalanche of work

In a 1927 publication, Hirth outlined steps farmers could take to help themselves by organizing creameries and promised to mail individuals step-by-step instructions.

…I think the farmers of any community should finance a Cooperative Creamery if as much as 500,000 lbs. of butterfat is produced annually within a radius of 15 or 20 miles of the center of such community—and with proper management such a Creamery should net the surrounding dairymen at least 5 cents per lb. more for their cream than they are receiving at the present time.

Already such plants are in the process of financing under the leadership of the M.F.A. at Springfield, Clinton, Chillicothe, Gallatin and Trenton, and by reading The Missouri Farmer you will be kept in touch with this splendid new movement which promises to place the dairymen of Missouri on a more profitable basis than ever before.

Hirth immersed himself throughout the late teens and until his death in 1940 with researching business plans, juggling suppliers and hiring contractors to construct MFA businesses. Hirth served as the unofficial secretary of the organization. All the while, Hirth, operating with conviction not salary, maintained his own pure-bred livestock farm, sold his livestock far and wide, edited The Missouri Farmer, sold print ads and hired individuals to sell automobile insurance throughout the 20s and 30s.

Organization money came in the form of MFA dues, which began as $2.50 per year. Of that sum, MFA kept 50 cents, plus a brokerage fee on commodities handled. No great sums were involved, however. In fact, as Ray Young reported in his book, "The earliest MFA financial statement reported a net gain for MFA [for Hirth's office use] in 1921 of $373.18, with a notation that it was later changed to $64.78." Membership dues were lowered during the depths of the Great Depression to $1.

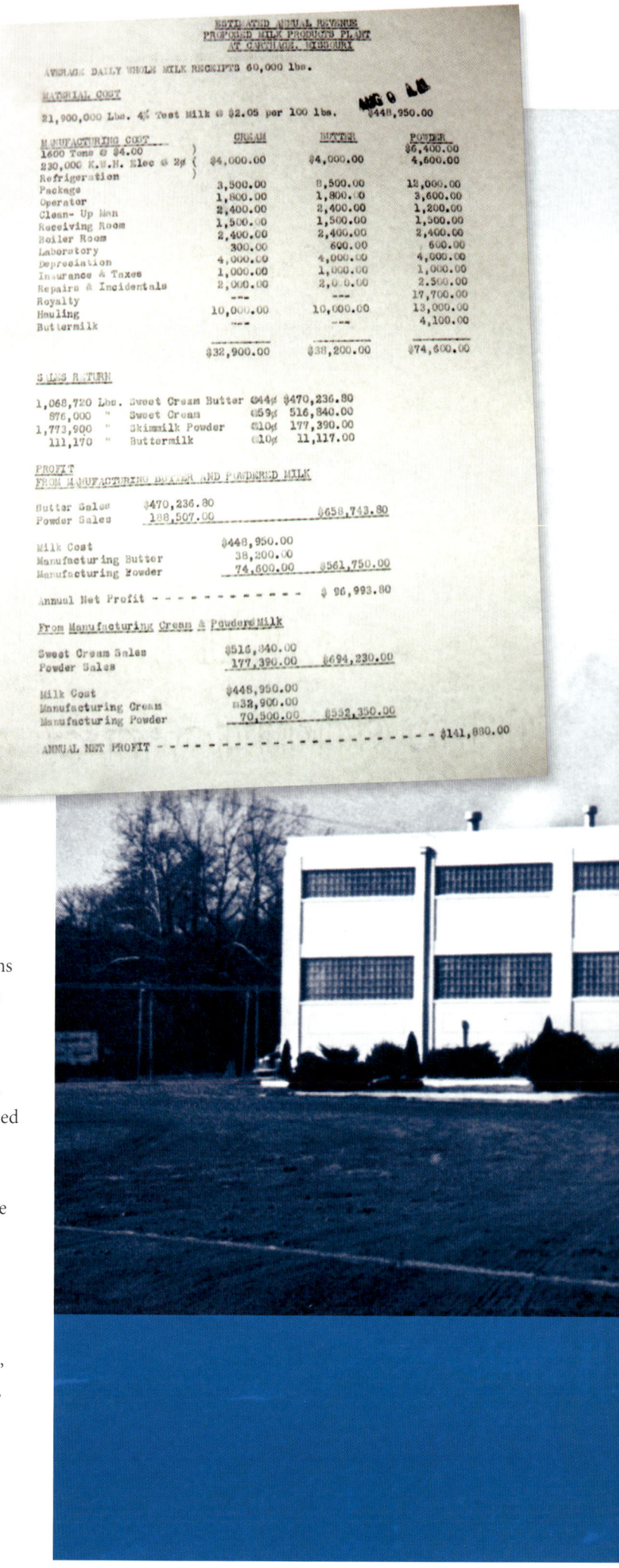

▲ MFA Producers Creameries sprouted all over the state with locations initially in Springfield, Clinton, Chillicothe, Gallatin and Trenton.

▶ **Opposite Page:** William Hirth developed detailed business plans for each of the proposed MFA facilities, estimating costs, revenues and potential profits. This business proposal for Carthage has four single-spaced pages of analysis. (Western Historical Manuscript Collection)

▲ Shelbina Producers Cold Storage shipped farmer products by rail to Chicago, Cincinnati, Canada and New York. Its early management faltered and Hirth was able to provide management control by the mid-1930s.

▼ The Ozarks were known as the Land of a Million Smiles when the MFA Producers Creamery began marketing products. Their dairy products carried the "Land O' Smiles" brand. Hirth double-checked with Land O' Lakes to make sure MFA was not infringing on its branding. (Western Historical Manuscript Collection)

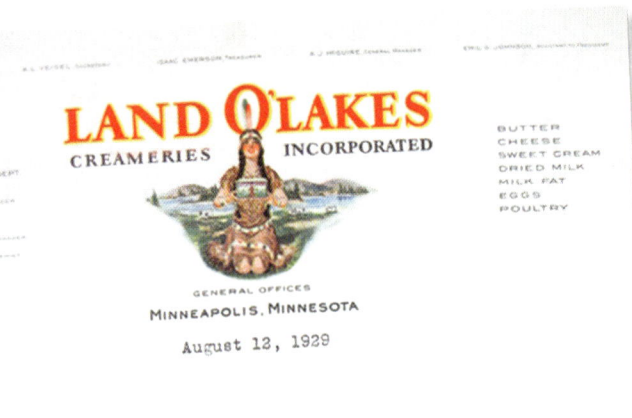

As Hirth wrote in 1939 in "The Romance of the Missouri Farmers Association," "How I ever bore up under the avalanche of work with which these activities deluged me seems a mystery to me now as I look back, for the demand upon me to make speeches was overwhelming, and all the orders for flour, feed, salt, coal, fertilizer, binder twine, etc., were handled through my office."

William Hirth was a religious man, but his workload would not allow Sunday as a day of rest. In a Feb. 6, 1932, reply to Ewing Y. Mitchell, state chairman of Franklin D. Roosevelt's presidential campaign, Hirth lays out a portion of his work schedule: "I regret having had to wire you that I could not meet you tomorrow, but usually my Sundays are filled with engagements at least a week ahead, for it is on this day that our fieldmen come in from different parts of the State to talk over the coming week's work."

The field men he referenced were employees paid to help organize farmers, explain the structure of exchanges and help develop larger organizations such as cold-storage businesses, elevators and like businesses. They also served to smooth troubled waters.

For instance, in 1927, with Hirth's prodding, Producers Creamery Company was incorporated and built in Springfield. Hirth's field men had a hand in that organization. Hirth, himself, was heavily involved in all aspects. Within a few years, the cooperative, affiliated with MFA but locally owned and managed with its own board of directors, became the largest creamery in the United States.

WILLIAM HIRTH AND THE BEGINNING OF MFA

To help individual farmers and farm clubs proceed, Hirth dealt with bank loan officials to arrange land, building and equipment purchases, but neither MFA, as an organization, nor Hirth outside of his advisory role had a direct hand in running the business. That doesn't mean Hirth was disinterested. Despite attempting a hands-off approach, in January 1932 correspondence with H.B. McDaniel, president of the Union National Bank of Springfield, Hirth took great pains to outline Producers Creamery's patronage rationale in view of the Depression, "considering the desperate conditions in which farmers find themselves."

Hirth was queasy about paying dividends and had consulted with W.T. Crighton who managed the creamery. "I asked Mr. Crighton the other day," wrote Hirth to McDaniel, "whether the payment of approximately $5,000 in dividends would seriously cripple his working capital, he said that it would not for the next several months, but that along in August, September and October he might need approximately $10,000 for big movements of sweet cream which always comes at this time of the year, and should he need a little help during this period I hope you will give it to him. Of course if in the meantime we can show fairly good net profits we may not need this help at all."

McDaniel by return mail hastened to assure Hirth that he had consulted with Crighton who had recommended against paying the $5,000. Wrote McDaniel, "We appreciate the business that is given us through your organizations, all of them being, so far as we are able to judge, well managed. I think they did the right thing in not declaring a dividend with business conditions as they are."

A very much nagged man

Hirth's role extended to personally soliciting and interviewing candidates to run the creamery. The Hirth Papers at the Western Historical Manuscript Collection provide a snapshot of the process (with a politically incorrect dose of humor). Presciently wanting the Springfield creamery to be the nation's biggest and best, Hirth was in the midst of conducting a national search for a manager. Applications flowed in. Hirth, a man with a penchant for follow-through, checked references.

Hirth wrote a friend at Clemson Agricultural College in South Carolina for a reference on one applicant. The Clemson professor's answer? The applicant was professionally competent but had baggage—his wife. "We always felt that [name withheld for any surviving relatives' sake] was a very much nagged man. I rather think that this would not interfere with his work in a commercial way."

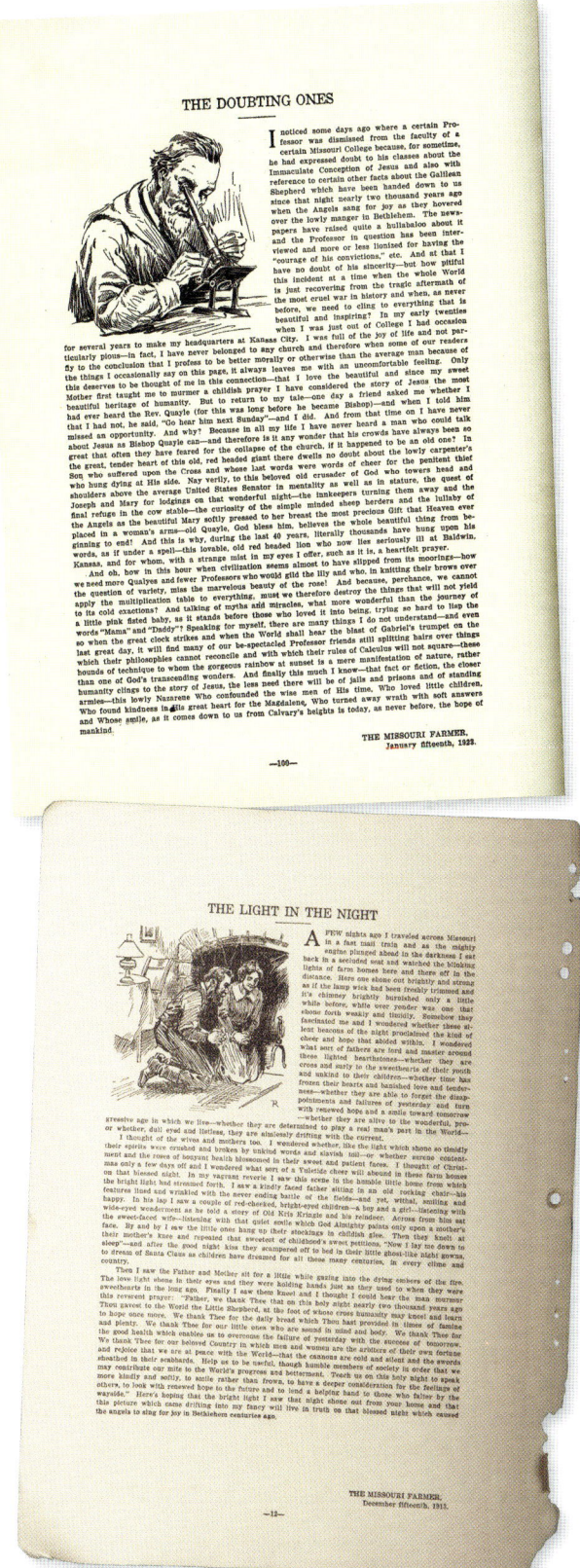

▲ William Hirth printed and published several books (called Afterthoughts) of his original essays. This one is as appropriate today as it was when it was printed.

▲ This article is reprinted from Afterthoughts. Hirth frequently wrote of farm life, encouraging farm families to improve not just production but their family lives as well.

Although Hirth did not hire the applicant for the Springfield creamery, he did hire him for a subsequent MFA creamery, which the man ran competently—with one caveat. Other Producers Creamery managers found him cold, distant and uncooperative. He tended, they complained, not to listen.

Hirth kept his hand in the business, consulting with Crighton on cheese making. Hirth wrote Crighton, "You ought to install the kind of show case that is used by an up to date grocery merchant in order that you may be able to handle the cheese just as such a merchant would and thus be in position to judge for yourself in what condition it would move to the consumer. In any case I once more repeat that we must be certain that the cheese will meet every expectation before we undertake to market it in a big way."

Crighton replied, "…I believe now we have perfected our cheese to meet the requirements of a most critical market. We have also received some encouraging reports from large concerns on the use of our protein and I feel we can look forward to getting some nice business in the East this coming season." Hirth simultaneously was writing J. Frank Grimes of the Independent Grocers' Association in Chicago: ". . . we will be in position to supply such American cheese as Chadder [in original], Twins, Flats, Daisies and Longhorns, and I therefore wish you would have your Purchasing Department to advise me whether they could use some of this cheese, and if so how much per week? Also whether you would want a full cream or partly skimmed cheese. Of course this is predicated on the idea that we will offer you high quality products on a competitive basis from private manufacturers."

Before the year was out, creameries were constructed in Clinton and Cabool. Over the next 10 years, Producers Creamery Company continued its expansion with creameries in Chillicothe, Emma, Kirksville and Popular Bluff. Again, management structures differed widely, but control, for the most part, remained at the local level, a structure that led almost immediately to trouble.

Providing a glimpse of the problems to come, in 1929, the head of an engineering company out of Chicago wrote Hirth on his company's plans to construct milk plants at select locations in the state, anticipating buyers.

Hirth rebuffed his proposal: "I have your letter of July 11th, and I want to take this occasion to say that the Missouri Farmers' Association will regard as unfriendly any activity on your part to finance Whole Milk Plants in this State, and your suggestion that we would be permitted to take over such plants later on does not change my view of this matter in the slightest degree." Hirth had expansion plans of his own in farmer-designated areas. He would use the engineering company, but he, not the engineers, would determine which locations best served farmer interests.

Further down in Hirth's reply is the nugget in his response to the engineer's suggestion that MFA had installed outdated technology in its existing plants. "The only point at which this has been done to my knowledge is at Cabool," Hirth wrote, "and this was without our consent, and only recently I suggested to our people at this point that they consider switching to your process in the not distant future." Note "without our consent" and "suggested to our people."

That hint of lack of control over facilities foreshadowed conflict to come. A.B. Drescher and C.L. Cuno, two early field men whose letters to Hirth survive in the Western Historical papers, constantly preached the necessity of MFA gaining more management control of newly created businesses.

Here is where the management structures first began to fracture. Hirth's early organizing successes included produce plants at various locations in the state. With Hirth working feverishly behind the scenes,

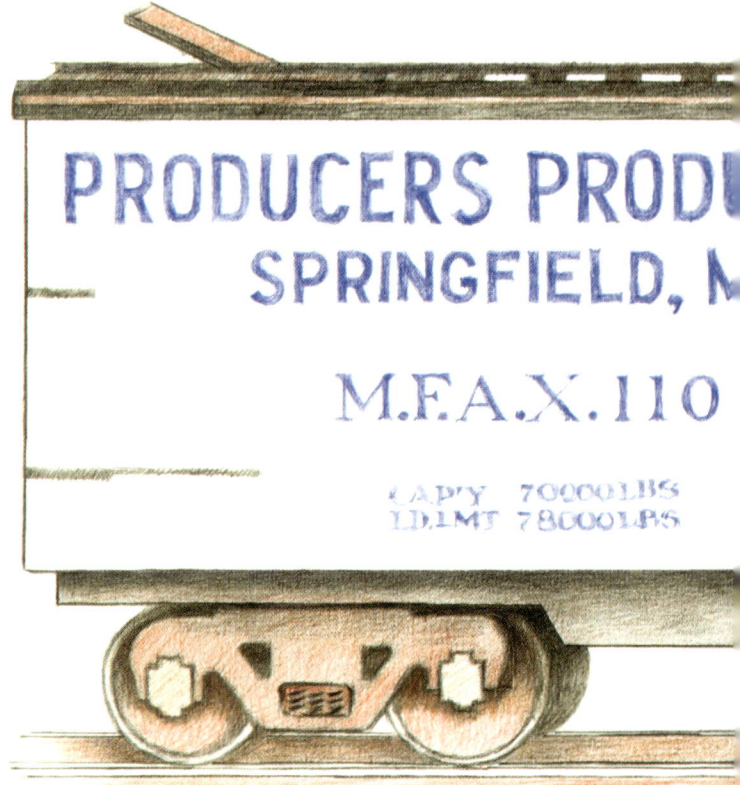

▲ Springfield Producers Produce is at the center of this impressive electricity grid. Two older photos in MFA's archives show this building began as Producers Cold Storage. In that earliest photograph, a truck at the loading dock has Buffalo Farmers Exchange on its door.

▼ By 1926, MFA Producers Produce companies were shipping 1,528 railcars of eggs and 936 cars of live and dressed poultry with a combined volume of $9.5 million. (Pat Hanna)

Proud Past, Bright Future: MFA Incorporated's First 100 Years

▲ Sedalia MFA Producers Produce Company caused Hirth nothing but headaches during its early years. Hirth had no direct control of the business, which teetered on bankruptcy until Hirth arranged professional management.

▼ Many of the more organized farm clubs even published their own newspapers. MFA activity at Springfield resulted in some of the biggest and best cooperative-owned businesses. Springfield would become known as the cooperative capital of the Midwest.

the first Producers Produce Company opened in Springfield in 1921 and was an immediate success.

Plants in Sedalia, Shelbina and Chillicothe quickly followed. By 1922, MFA Producers Produce plants had also opened in St. Joseph, Medill, St. James and Clinton. All continued the practice of local control. The Missouri Farmers Association board was given the opportunity to appoint several board members. But the MFA-appointed board members did not have a majority. Hirth immediately came to regret this structural problem.

In the main, the produce companies collected poultry and eggs from area MFA exchanges and distributed the products to Cincinnati, Chicago, New York and Canada. As Ray Young describes in his book, by 1926 these plants were shipping 1,528 railcars of eggs and 936 cars of live and dressed poultry with a combined volume of $9.5 million. Those totals led the nation in the value of poultry products shipped cooperatively.

Despite these early successes, the businesses soon became administrative nightmares. Springfield-based Producers Produce was run by

32   WILLIAM HIRTH AND THE BEGINNING OF MFA

Hirth's confidante (and hand-selected) Lee Farnham, a highly competent businessman. But the other plants quickly descended into chaos.

Sedalia's plant caused Hirth constant headaches, in the main, because his advice was rebuffed or ignored by the management and local owners despite Hirth's inside knowledge of the workings and structure of the Springfield plant.

Relationships with area MFA exchanges were critical to success. For instance, if exchange managers didn't like the local management of an individual MFA business, they would withhold their support. At Sedalia, the Producers Produce Company constantly teetered on bankruptcy.

The indispensable Drescher wrote Hirth earlier in the year warning of feuding and factionalism in the Sedalia area. "The fact is that either the Exchanges in the Sedalia territory MUST SIGN UP 100% and BE THAT or the Sedalia Central plant will go UNDER—no question about that."

Hirth, after personally beseeching the board, wrote Drescher, urging his personal intervention: "The Board voted to invite Springfield to supervise the Sedalia plant, and a joint conference will be held during the next few days, and I think you should be present. I made it plain to Mr. Steeples [the manager at the Sedalia plant] and the Board that Mr. Farnham would not take charge unless he has a free hand, and Mr. Steeples expressed his entire willingness to place himself subject to Mr. Farnham's ideas and policies." It took looming bankruptcy to force that concession.

These collisions with bad management and near misses with bankruptcy were opening Hirth's eyes to his mistakes in management structure. By 1929, Hirth was asking members to inform him quickly if their elevator, exchange, livestock shipping association or other MFA agency was "sick." Remember, he said,

while we are always glad to hear about our successful agencies, **we are ten times as anxious to hear from an agency that is in trouble!** [Emphasis in original] This is one reason why I am so deeply interested in putting our Regional Auditor's Plan into effect, for then our weak plants can be kept under constant surveillance.

But for the time being, if your agency is in financial danger, or if it is losing volume, get these facts to us without delay, and then we will see if our Association cannot do something to help out. In my time I have seen a lot of M.F.A. agencies drift close to the rocks, then right themselves, and today these agencies are among the most successful we have, and therefore because the skies are clouded today doesn't mean that the sun will not shine tomorrow. But when an Elevator or Exchange is headed down hill, the Association is entitled to know it, for if it should close its doors, then both the Association and the nearby agencies must bear their full share of the disaster.

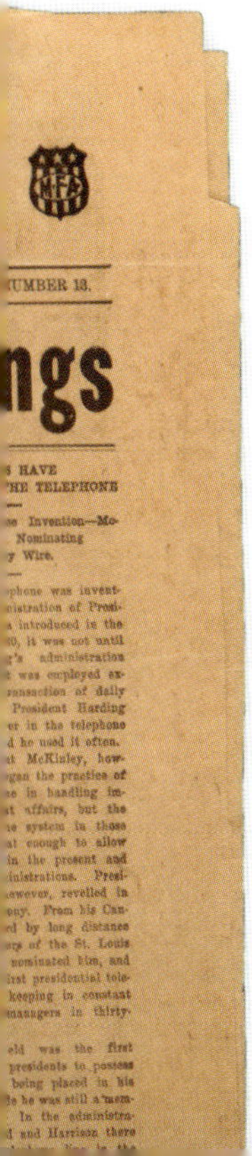

MFA Stores

Brookfield

Brunswick

Cabool

Cairo

California

The right man for the job

Hirth learned important, but not complete, lessons from these early enterprises. Previously, in the early 1920s, responding to farmer demand, MFA exchanges were clamoring for mixed feed. Local MFA groups organized the Farm Club Mill and Feed Company of Springfield. The enterprise got off to a rocky start. By 1923, the MFA organization had loaned the mill $3,000. Still, the business floundered. Hirth, through MFA, advanced another $2,000 and changed the name to MFA Purchasing Department and finally MFA Milling Company.

Two things were needed immediately: a new facility to meet growing demand and a corresponding influx of capital. Mill officials identified a milling property as a likely solution. Hirth called on friends and political connections in the Springfield area, asking Ewing Y. Mitchell, a prominent lawyer in Springfield, how best to obtain a $50,000 loan. In documents supporting the loan request, the mill is described as a property of "brick, inside the floors are wood construction. The elevator is a crib elevator with sheet iron on the sides, and has a capacity of 125,000 bushel. This elevator was built in 1908 and is practically as good as new."

Capacity for 20 railcars was available with switching tracks served by the Frisco and the Missouri Pacific. MFA received the loan and incorporated the business under the Non-Stock Cooperative Act. This time a majority of MFA board members populated the board of directors of the MFA Milling Company.

Still, the business languished. On Jan. 20, 1932, Hirth received a letter from MFA and MFA Milling Company board member T.M. Chapman of Ozark. "…I cannot refrain from mentioning some things that are playing havoc with our business that you may not be familiar with. In the least about everything has happened that could be thought of to handicap the Mill at Springfield. In the first place competition has been very strong in our line. The low prices for poultry and dairy products has left not much incentive for buying high-class mixed feeds in order to increase production."

Chapman goes on to complain of MFA exchanges using their own mills or grinders

MFA Milling Company of Springfield blossomed into a feed powerhouse and remained a force for decades. As the above sack demonstrates, seed was also a product.

"hooked on to ford cars or small tractors that move from farm to farm and grind everything at a very low cost. …The thought of the Mill going down gives me a nightmare. I can't think of anything that would be more humiliating to me than to admit that the Mill is a failure."

Hirth knew better. Despite a majority MFA board oversight, the problem was management. And though it took until 1935 to find the right man for the job, Hirth finally tapped John F. Johnson, an MFA auditor and exchange manager at first Marshfield and then West Plains. Johnson proceeded to build the business from 354,778 bags of feed to 10,566,446 hundred-pound bags by 1959.

The MFA Milling Company was off to a great start and a greater history. For the nearly 40 years of Johnson's leadership, MFA Milling Company would dominate the feed business in southern Missouri and northern Arkansas. MFA Milling Company would become one of the largest cooperative feed businesses in the country, serving farmers in Arkansas, Kansas, Louisiana, Mississippi, Missouri, Oklahoma, Tennessee and Texas and returning more than $26 million in cash patronage refunds.

"We'll buy the whiskey."

With the Producers Produce problem solved at Sedalia, another management disaster brewed on the horizon: Chillicothe Producers Produce Company. Drescher had been assigned territory that included Chillicothe. His reports concluded the manager was running a good-ole-boy business, paying higher prices to friends, buying outside the system and delivering inferior products that were seriously undermining MFA's reputation.

At the time, MFA through the Producers Produce Companies was providing one-fifth of the U.S. eggs exported to Canada. One Canadian shipment included 37 railcars of MFA producers' eggs. A tariff enacted by the United States Congress ended those shipments despite Hirth's vocal, and public, complaints.

In terms of profitability, Chillicothe Producers was 75 cents per egg case below Springfield which equated to $300 per railroad car. Drescher had no luck with Fred G. Peters, manager of the Chillicothe company. In fact, communication with Peters was fractious and confrontational.

Drescher was disgusted with what he saw as inept management. "The showing at Chillicothe is just about as it has been for two or three years—did it ever occur to you that with a volume of over two million and paying LESS than Springfield pays and this nearly every day in the year, the profits are almost nothing—I wonder if the wise men on the Chillicothe Board ever think of that, or of ANYTHING."

More to the point, Drescher noted Chillicothe's business practices were damaging MFA's brand: "I wrote you what the concern at Cincinnati said or wrote about MFA poultry, they said they would not be interested in it AT ANY PRICE—this because of their experiences with Chillicothe."

Drescher hammered out letter after letter apprising Hirth of the situation, which Drescher felt, was deteriorating by the hour. His solution?

MFA was supplying eggs by the railcar to cities from Missouri east. To make certain MFA products were in demand, Hirth urged members to command premiums in the central markets by improving egg quality. His article ended with: "I wish you men would take this matter up with your wives…" (Western Historical Manuscript Collection)

Centralized management of the MFA companies. "I think the time for ACTION is right here, by action I mean this oft talked about SHOW DOWN—not any d-n audit of the books of Chillicothe or Shelbina or any other point as we had that sort of thing once and the Board at Chillicothe never batted an eye—but what I mean is to either BUY or lease or take over in some way the BUSINESS of the MFA—its Central plants FIRST, especially those in north Missouri, and CLEAN HOUSE . . .." Drescher wanted management and boards fired and replaced. More to the point, he wanted centralized control. He wanted Hirth to: "START OVER AGAIN and start RIGHT."

Furthermore, Drescher wanted to hire three individuals who were outstanding managers who knew the business thoroughly, divide the state geographically and put these men in charge of the poultry and egg plants. The three would report directly to Hirth. Drescher's prescient plan mirrors MFA's current corporate structure.

Hirth and Drescher were as one on the idea of central control. Hirth made no secret of his desire to replace management and the board or start a new facility nearby. He had ample evidence that something was amiss in Chillicothe.

For two years, he had been demanding an audit of the company's books, something he could only do with permission. Peters had refused to open his books, raising Hirth's suspicions and underscoring his desire for control. Hirth was also receiving inquiries from members.

O.A. Grim of Trenton wrote Hirth asking about the situation. Grim's letter is not in the Western Historical Manuscript files, but Hirth's response is. "I have your letter with reference to the situation at Chillicothe, and apparently things are in bad shape down there. However you will appreciate that our Association cannot interfere until the Exchanges up there ask us to do so, for if we took a hand on our own initiative we would be charged of trying to make a bad situation worse."

By fall of 1929, the situation became untenable. In mid-October, Hirth and Drescher attended a meeting of the Chillicothe Producers Produce Company. Management and board members, as well as a large contingent of local members, expected Hirth's attendance and were loaded for bear. Manager Peters and Board Member Frank Scott were anti-Hirth and rallied others to their cause. Scott had been elected secretary of MFA but had been dismissed in the past year. Like most of the renegade locations, they had no desire for what they termed "outside interference."

Peters began the meeting by delivering his management report. Hirth rose to question him. Board members immediately pounced, turning the meeting into an inquest on Hirth's character, accusing him of using The Missouri Farmer to further his own cause and further accusing him of buying liquor for a commission company distributing MFA goods.

That last part was true, Hirth replied. "If it takes whiskey to help the cause of cooperation, we'll buy the whiskey." The situation deteriorated. The audience enthusiastically re-elected their board, effectively repudiating Hirth who had brought a list of replacement candidates. Hirth went home determined, not defeated. He understood the implications. Furthermore, he understood just what was at play: his adversarial relationship with Howard Cowden, Frank Scott and the newly formed Federal Farm Board.

"START OVER AGAIN and start RIGHT."

—A.B. Drescher

M·F·A
ALFALFA TRIALS

★★ Chapter 2 ★★
Convergence—Politics, Federal Farm Board, Cowden, Farm Bureau and Paybacks

"Facts, hell."

The newly created Missouri Farmers Association needed more than organizing; it required political capital both state and national. William Hirth was uniquely suited for that job, too. By the mid-1920s he was a member of the executive committee of the American Council of Agriculture, representing MFA. More importantly, on the national scene in 1925, the Corn Belt Committee was organized with a membership drawing together the largest group of influential farm leaders in the country.

The Corn Belt Committee represented 24 farm groups including Farm Bureau, Farmers Unions, the Equity, the Grange and multiple Corn Growers Association groups as well as many others. The group's purpose was to speak on agricultural legislation as one, make agriculture's positions known to legislators and the public and pronounce "upon all matters concerning agriculture." Hirth's oratory skills were well known and respected by the group. His national standing made him a natural leader. He was selected president of the newly formed Corn Belt Committee and retained that position for several years. In his capacity as chairman, Hirth helped make agriculture a national concern. His speeches and activities also attracted the attention of national leaders, including one who would soon be mulling a presidential bid—Franklin Delano Roosevelt, governor of New York.

Hirth held that national attention until his death. And with good cause. Because of his national leadership efforts and accomplishments, Hirth's name would be formally submitted unanimously by the entire Missouri delegation—Democrats and Republicans—to President Calvin Coolidge for the position of secretary of agriculture. Hirth demurred, asking that his name be withdrawn. His appointment, said Hirth, could only mean "hell in the kitchen rather than the harmony which the President undoubtedly desires and has a right to expect."

But Hirth's focus was not simply national. Missouri had plenty of obstacles for farmers trying to grow and move products. Country roads were impassible for large chunks of the year making it difficult if not impossible to deliver farm products to towns and MFA markets. Solution? Hirth jumped with both feet into initiating legislation authorizing farm-to-market roads and personally lobbied political leaders and newspaper editorial boards. Through his highly influential The Missouri Farmer, he chronicled his efforts and urged the membership to lobby alongside.

"Many times during recent years the farmers of North Missouri were hardly able to bury their loved ones because of bad roads," intoned Hirth in a speech at the MFA convention, "while their children are forced to wade ankle deep through mud in going back and forth to school, and so bad roads often keep our few remaining rural churches closed for weeks at a time." He convinced the organization and its board to get behind legislation pulling farmers out of the mud.

Road legislation soon passed at the state level. In the long term, the legislation failed to achieve all Hirth demanded because of what Hirth described as a politically controlled Highway Administration.

The Missouri Farmer, he said in 1928, "is unalterably opposed to any more bond issues for cross-state highways, until something substantial has been done for the neglected dirt roads." Hirth carried on his farm-to-market road campaign until his death in 1940. His successor would deliver Hirth's promised improvements.

The state/national/agricultural focus took much of Hirth's already busy schedule. In his Jan. 25, 1925, talk with managers, Hirth gave a glimpse of his Washington efforts: "…I am just back from Washington, where as a member of the Executive Committee of the American Council of Agriculture I appeared before the President's Agricultural Commission and also before the Agricultural Committees of the House and Senate—and the latter was a 'showdown' such as perhaps these Committees have never witnessed before."

Hirth was famous for his political "showdowns," at one point locking the door to the Agriculture Committee room so that no member could leave until he finished. In the early 1930s, Hirth testified before the Senate Agriculture Committee.

Hirth, by contemporary accounts, was a large man for his times, had bushy eyebrows and cast a commanding presence. He finished his testimony, turned on his heel and stalked out of the room. U.S. Secretary of Agriculture Henry Wallace, who had attended, chased after Hirth, finally catching him in the Senate corridor. "Mr. Hirth," he is reported to have said, "Mr. Hirth," he repeated when Hirth stopped and turned to face Wallace. "You didn't give that committee some of the facts." Hirth scrunched down his eyebrows, scowled at the U.S. Secretary of Agriculture and said, "Facts? Facts, hell. That Committee didn't need facts. They needed their minds changed."

And Hirth was up to the challenge of changing minds. Testifying to Hirth's power of persuasion, A.W. Ricker, secretary of the Corn Belt Committee, described Hirth's speech to members of the Corn Belt Committee immediately following President Coolidge's first veto of the McNary-Haugen bill, a piece of legislation listed by historians as undoubtedly Hirth's crowning legislative initiative. "Hirth unconsciously revealed himself as a giant of purpose, intellect and courage," said Ricker. The speech drew national attention.

By way of background, McNary-Haugen was an agricultural relief bill. It was vetoed by President Coolidge both times it was submitted. It also underscored why Hirth would have been a bad fit as Coolidge's secretary of agriculture. Nevertheless, historians applaud Hirth for his tenacity and congressional influence. That he could force the bill through Congress twice in the face of Coolidge's staunch opposition showed Hirth's influence, oratorical skills and political acumen in Congress.

In The Missouri Farmer Hirth explained his motivation for his famous speech after the first presidential veto:

> At the conclusion of the address…(which was approximately of an hour's length) I was importuned from all sides for copies of it—but since I had spoken entirely extemporaneously, and as the spirit moved me, this was impossible. …
> The occasion was a solemn one—through the President's veto message, the men who sat before me had been dealt a terrific blow, and the following 24 hours were to determine whether we were to gird our loins anew, or whether cowed, discouraged, and beaten, we were to run up the white flag and abandon agriculture to its fate. And therefore,

"Facts? Facts, hell. That Committee didn't need facts. They needed their minds changed."

—William Hirth

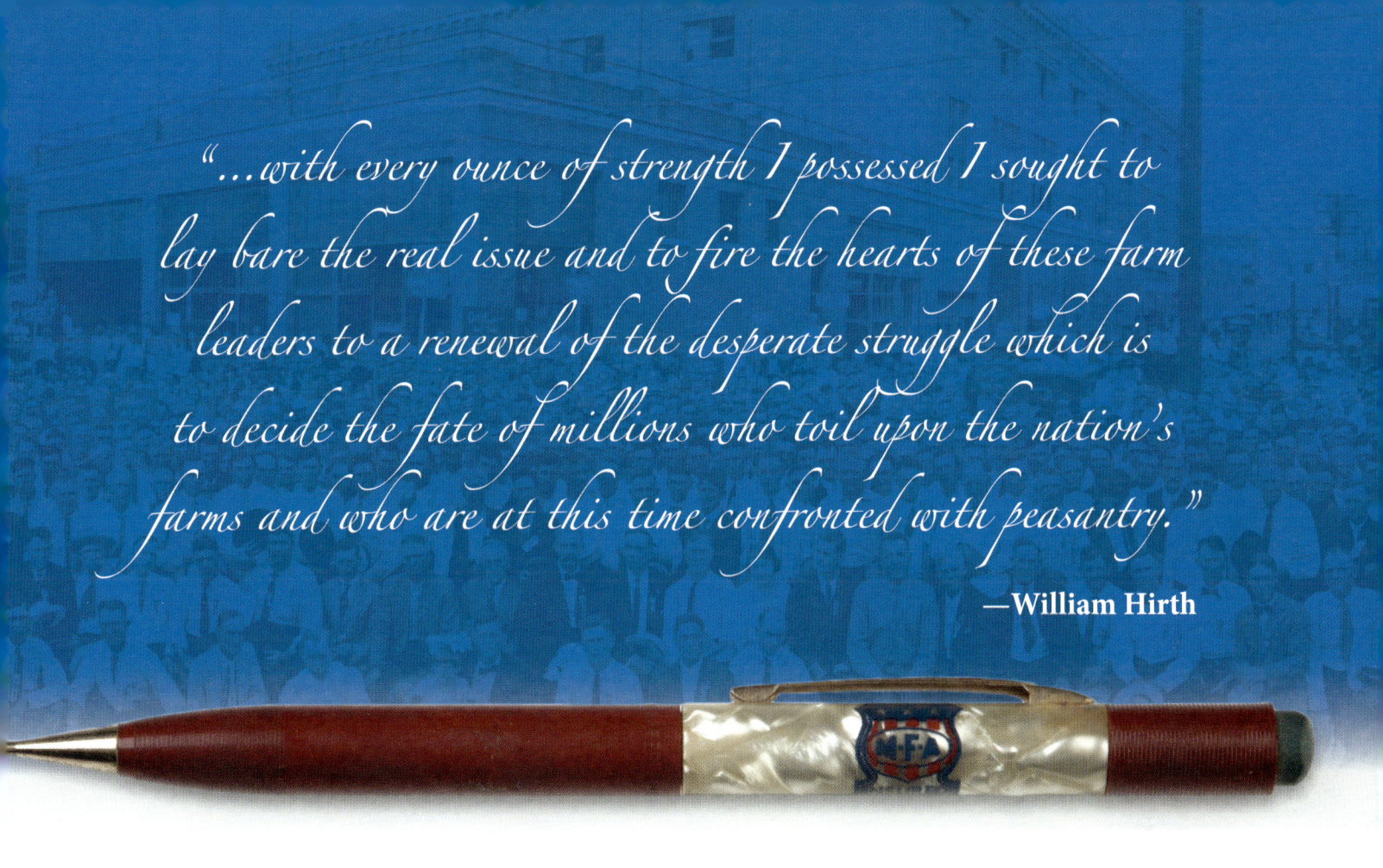

> "...with every ounce of strength I possessed I sought to lay bare the real issue and to fire the hearts of these farm leaders to a renewal of the desperate struggle which is to decide the fate of millions who toil upon the nation's farms and who are at this time confronted with peasantry."
>
> —William Hirth

if I spoke with more than my accustomed depth of feeling, and if the address is entitled to the estimate which some have placed upon it, there was ample reason—with every ounce of strength I possessed I sought to lay bare the real issue and to fire the hearts of these farm leaders to a renewal of the desperate struggle which is to decide the fate of millions who toil upon the nation's farms and who are at this time confronted with peasantry. And if my words achieved their purpose, it was because of the justice of the great cause for which I plead, rather than because of any of the arts of oratory.

The federal fiasco

At the precipice of the Depression, Hirth was actively lobbying the federal government for facility loans. After the Depression hit, Hirth stepped up his correspondence and lobbying as banks failed. Without the ability to immediately deliver capital, structurally sound businesses faced disaster. MFA had a large number of businesses. An even larger number of banks were failing. And as they failed, creditors were calling loans.

A Federal Farm Board had been formed to review loans for the express purpose of providing federal loans to businesses in just these situations. At the outset, Hirth had not been a fan. His preferred solution was a practical one: use existing Intermediate Credit Banks. In fact, consider his June 2, 1929, telegram reply to a request from U.S. Representative Clarence Cannon of Missouri: "I would contemptuously reject place on Federal Farm Board if it was offered."

Although Hirth was watching banks across Missouri and the country topple, he had no patience with new Washington bureaucracies and political theorists. Above all, Hirth was a practical man. He corresponded with his friend Senator Peter Norbeck of South Dakota (chairman of the Senate Banking and Currency Committee and patron of Mount Rushmore) concerning the newly organized Federal Farm Board and comments by its political creators.

Those individuals had suggested supplying money to farmers to help them federate, consolidate and build membership in cooperatives. Hirth found their designs "wholly unsound." What Hirth wanted was a ready source of capital for cooperatives already organized and able to offer sound collateral to back those loans. As he went on to explain, "I have never had the slightest patience with the idea that it is a legitimate obligation of the

Government to supply money to organize farmers, however distressing their plight may be; if such a proposal were made with reference to working men and women, I think it would be considered preposterous, and I see no reason why it is less so when it applies to farmers, even admitting their rather pitiful economic condition at the present time. In other words, unless farmers still have enough red blood in their veins to organize themselves together for mutual benefit, then we might as well let the tail go with the hide."

Hirth had been in constant communication with his friend Arthur Hyde, who was the current U.S. Secretary of Agriculture under Coolidge. Hyde had been governor of Missouri (1921–25) and he and Hirth had long-standing ties. Hirth didn't seek largess; he envisioned the federal government supplying loans only to those concerns with sound balance sheets and with the loan offset by proper security. He wanted politics out of the equation. But separating politics from federal programs was too much to expect.

Hirth moved back and forth between his friend Agriculture Secretary Hyde and Alexander Legge, chairman of the Federal Farm Board. Hirth stated in his letter to Legge, "…considering that the M.F.A. is almost the biggest cooperative in point of operations in the Country, surely I am not asking for anything that is unreasonable, and remember we will be glad to establish the soundness of every loan for which we ask, and in very few instances will we ask for more than 50% of the fair physical value of the agency in question."

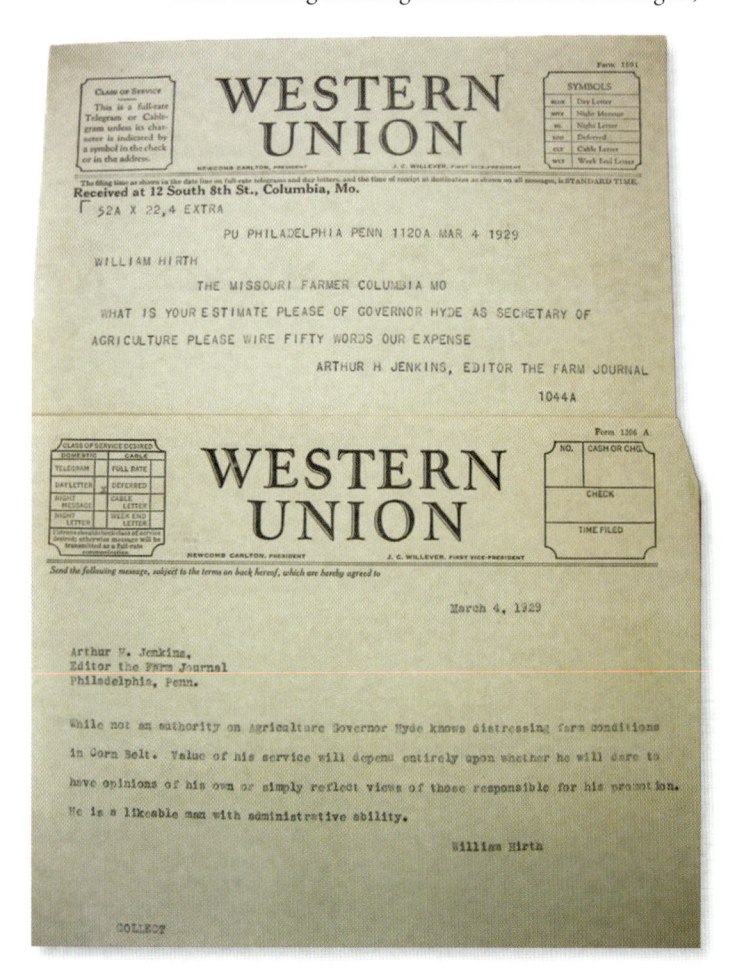

> "In other words, unless farmers still have enough red blood in their veins to organize themselves together for mutual benefit, then we might as well let the tail go with the hide."
> —William Hirth

Missouri Governor Arthur Hyde and William Hirth were friends. Hyde was President Hoover's choice for U.S. Secretary of Agriculture. As Hirth would write Hyde, "I have sufficiently recovered from the shock of your appointment so that I am once more able to sit up and take nourishment. Meanwhile, I send you my heartiest congratulations." Hirth went on to explain, "…my interest in Agriculture is so deep that it transcends friendship." Hirth would be a frequent critic of Hoover's (and Hyde's) policies. (Western Historical Manuscript Collection)

Hirth soon stumbled across an unforeseen obstacle. As Hirth discovered, Fred G. Peters of MFA's Chillicothe Producers Produce had been in communication with Legge as well. Peters wanted to sever all ties with MFA. He wanted direct loans to Chillicothe with no interference from the MFA organization. And he wanted a seat on the Federal Farm Board to put a thumb in Hirth's eye.

Throughout the start of the Depression, Hirth had been doing the legwork for all of MFA's various businesses. His intent was for MFA, as the presiding organization, to receive loans for whichever MFA organization was most in need. From a management standpoint, the practice was sound. It would also provide him with leverage in dealing with errant MFA organizations. Legge called the requests for loans "unnecessary duplication" and wanted the matter "to be cleared up."

Hirth was undeterred, writing Dec. 31, 1929, to Legge: "I have your letter of December 28th, and there is really nothing further that I can say with reference to the Missouri situation—there is no chance for a compromise in these premises, it is purely a question of whether your Board will insist on doing business through the Officials of an Association, or whether it will recognize disgruntled factions within it."

In further correspondence to Legge, Hirth reflects the Depression's reality: "I just wanted to drop you a line to say that another bank has failed at Gorin, Missouri during the present week, and this makes five closed banks in Scotland County during the present year—the point is that each of these failures has not only seriously hurt our Farmers Exchanges in that County, but under the circumstances you can appreciate the almost impossible task we are up against when we seek to finance a Creamery or agency of similar size."

The loans never materialized.

Howard Cowden made no small plans

Howard A. Cowden was an able, ambitious young man in Polk County. In 1919 he hired on as secretary at Polk County Farmers Association, headquartered in Bolivar. The job of secretary at any of the farm clubs was new and expanding. As it developed, the county secretaries took orders from farm clubs in the county, pooled those orders and sent them to Hirth. Hirth, in turn, pooled county orders and arranged carload deliveries at volume discounts.

In The Missouri Farmer, Hirth asked for patience for the individuals in these roles. "Every one of them is 'plowing a new furrow,' and it will take them a year or two to hit their pace…in the past, too many farmers have wanted to put six bits into the till and pull out a dollar and a half the next morning. Remember, that in a large measure these county secretaries are going to have to create their own duties. …"

Hirth's calls for patience were unnecessary in Cowden's case. Cowden shone immediately. Right away, he caught the eye of Hirth who was effusive with his praise. Hirth saw great things in Cowden's future. And great things would come. But not at MFA.

At the time, Hirth was serving as unofficial overall secretary of the expanding MFA. He was MFA's chief organizer and event speaker. He was also publishing his magazine and actively leading on the state and national political scenes. His time was at a premium. He needed relief. He soon looked to the

MFA Stores

Canton

Carthage

Catawissa

Centralia

efficient and effective young man at the Polk County Farmers Association.

Cowden was quiet, thoughtful and soft-spoken. He was also aggressive, fiercely competitive and, as you'll see, manipulative. Hirth was impressed by and approved of Cowden's work. Cowden became a field man in short order and traveled the region, signing up thousands of new MFA members.

Hirth rewarded Cowden's early success by hiring him as the first official secretary of MFA. Cowden began his duties on the first day of 1922. According to Gilbert Fite in his history of Farmland Industries, when Cowden took over Hirth's duties as secretary, he found "little order in the affairs of MFA and little of anything that resembled a business system." It was fair criticism. With Hirth's workload, disarray of paperwork could be expected.

Cowden set to organizing the paperwork and implementing more structured business practices. He also enjoyed an almost unfettered luxury denied Hirth. He focused only on the business side of the evolving cooperative. He also took control of the handful of field men who were currently organizing farmers, farm clubs, exchanges and assorted MFA-affiliated businesses.

Hirth was delighted with the arrangement. Up to that point, Hirth had been working 16-hour days, six and one-half days a week—all without official position and no salary. Cowden's initiative and efficiency allowed Hirth time to devote to his other duties.

Specifically, Hirth fleshed out an idea to contractually bind farmers to deliver products to MFA facilities. Those contracts would guarantee a predetermined stream of grain, poultry, livestock and milk products. With those guaranteed volumes, MFA would strengthen its position (and by extension, the farmer's position) in negotiations with buyers.

The Producers Contract was adopted by the delegates at MFA's annual convention in 1923. William Hirth's concept for the contract was to end the practice of farmers dumping commodities on a down market, further depressing prices. From Hirth's perspective, locking farmers into a contract would allow MFA a strategic and orderly control of the marketing of those commodities. The first contract was signed by Aaron Bachtel whose farm club had proved the genesis of MFA. Each farmer paid a $10 fee to participate. According to Hirth, the purpose of the contract was "to pool the selling of farm commodities so they can be disposed of through great central Selling Agencies and thus place in the organized farmer's hands the power to have something to say about what he shall receive from the fruits of his yearly toil."

MFA's Producers Contract was an ill-fated attempt to contractually build large volumes of products for more marketplace power in negotiating with packers, wholesalers and other businesses. It was adopted by the delegates at the 1923 convention.

Elsewhere in the country, the contract idea was already successful in individual commodities such as tobacco, citrus, wheat and cotton. MFA's attempt was the first to lock in a contract across multiple lines of products.

Cowden was given control of organizing the contract sign-up. Hirth, mostly in Washington working on national legislation, took a back seat and offered advice. Cowden aggressively marketed the contract proposal, his way. MFA field men pushed the contract in the countryside. Simultaneously, Hirth pushed for the creation of an "Old Guard." To join this exclusive MFA club, members had to have brought in 10 new members in one year, organized one new farm club of at least 10 members or obtained contract signatures of 75 percent of farmers in a school district.

Hirth's advice to Cowden was to have the "Old Guard," as proven loyal members (men and women), recruit farmers to sign the contract and stay at it until 75 percent of farmers signed up in a county. Cowden's plan was to employ field men to gather as many signatures as quickly as possible and move to the next county regardless of the percentage of farmers signed. Cowden defied Hirth and implemented his own idea, employing the field men who reported to him personally. Hirth was constantly on the move. Cowden stayed put. Under Cowden's method, some counties barely reached 30 percent.

The contract never lived up to Hirth's expectation. Noting the effects of the Depression, MFA officially ended the contract in 1933. MFA returned all fees with 6 percent interest, thereby establishing even more credibility for the cooperative. Approximately 50,000 farmers had signed the contract, but not all had paid. Refunds amounted to $487,240 plus interest of $24,500.

Cowden continued to build the business and cultivate field men and, more importantly, board directors. Cowden was astute. He thought big picture. He saw opportunity and he moved forward in efforts to shape MFA into a more efficient organization. More to the point, Cowden kept his eye on the developing farmer need of petroleum products like kerosene, gasoline, oil and lubricants.

Henry Ford had begun mass production of the Model T, and Americans were beginning a love affair still in place today. Farmers were no different, with one exception. They also fell in love with tractors. And Henry Ford, himself a farm boy, designed and mass produced the Fordson Model F. It was the first small, lightweight tractor on the market. Better still, it was priced within a farmer's reach. Quickly capturing a 70 percent market share, the Fordson begat an agricultural quest for fuels and lubricants. By 1925, Ford had manufactured half a million Fordsons.

Cowden saw potential and quietly began a quest of his own that would lead him to fortune and to trouble.

▲ To become a member of MFA's Old Guard, a man must sign up at least 10 new members, organize a new farm club of 10 or more members, obtain signatures of 75 percent of farmers in his school district to the MFA contract or drive an MFA field man for at least one week. The Old Guard grew to more than 3,000 members in a short time.

▲ Modeled after the men's Old Guard medal, the Home Guard medal was presented to members of the Women's Progressive Farmers Association. To earn the medal, women had specified criteria to meet, mainly in terms of organizing clubs and enlisting members. This medal was donated by Gene Murphy.

Proud Past, Bright Future: MFA Incorporated's First 100 Years 45

Farm Bureau, consolidation attempts and the peace treaty

Today, Farm Bureau and MFA share common goals and philosophies, participate jointly in projects and cater to many of the same members. That wasn't always the case. Farm Bureau and MFA were both created in the same timeframe. MFA, however, was the first to provide immediate economic benefits.

Farm Bureau was more or less a creation of Extension. As James Rhodes, professor emeritus of agricultural economics at the University of Missouri, points out in his analysis, Missouri legislation permitted county governments to sponsor and finance county agents, whose funding came courtesy of the passage of national legislation. The early Farm Bureau remained closely tied to Extension, USDA and agricultural colleges.

From Hirth's perspective, he bore no ill will toward Farm Bureau, but considered "that under the Smith-Lever Act which brought it into being its function is purely educational, while on the other hand, the M.F.A. grapples first hand with the farmer's marketing problems, and admitting that the educational field is vital, in this hour when the average farmer is struggling desperately to pay his interest and taxes, **is not a better price for what he produces, and a lower price on feed, flour, fertilizer, etc., at least equally vital?** [Emphasis in original]"

Hirth saw unnecessary duplication in terms of associations. From his perspective, he had organized farmers first. What's more, Hirth saw Extension as an MFA roadblock. He thought correctly that Extension had chosen sides. Hirth was anti-Extension but pro-University of Missouri.

As he wrote in The Missouri Farmer, "And while, on the other hand we protest against lavish expenditure of money for a swarm of Extension men who will not be needed from this time forward, we bear cheerful testimony to the fact that such men as Dr. Whitten, Dr. Connaway, Prof. Eckles, Prof. Haseman, Prof. Kempster, Prof. Weaver, Prof. Allison, Prof. Miller, Prof. Trowbridge, Prof. Etheridge, Prof. Talbert and others have rendered a service in recent years which the farmer of this state can never repay. . ." Many of the articles in The Missouri Farmer were written by these professors and others. Hirth considered it MFA's mission to pass along to farmers the research and expertise provided by these professors.

But claimed or not, similarities were great between Farm Bureau and MFA, so much so that many saw destiny in combining the two into one entity. In the late teens and early twenties, several attempts were made to do just that. As late as the 1940s, merger proposals bubbled up.

At the start, Hirth was having none of it. He saw Farm Bureau as an association of farmers united to provide agricultural education and structured such that it could at some point provide collective buying and marketing. MFA had already plowed that ground, he said. MFA was designed for economic reasons, first and foremost, said Hirth. Besides, as noted above, MFA had hundreds of farm clubs, hundreds of exchanges, dozens of elevators and a large number of economically viable agricultural businesses serving a national market.

Hirth envisioned Farm Bureau's 1921 overture as an attempt to assume control of "our hundreds of elevators and Exchanges and larger marketing agencies in whose investment they didn't have a single dollar." Hirth's rebuttal to Farm Bureau's outline of cooperation was termed a peace treaty.

The peace treaty stated Farm Bureau would forgo cooperative marketing and purchasing, something Farm Bureau dismissed out of hand. According to Vera Busiek Schuttler's "History of the Missouri Farm Bureau Federation," one of the Farm Bureau officials on receipt of the treaty sarcastically referred to MFA's proposal as "Welcome to the MFA." The merger talks dropped from public sight—or at least Hirth's sight—for several years.

Still, Hirth was not unalterably opposed to a merger between the two organizations. He stated publicly and repeatedly that "under the proper conditions I will welcome it most cordially, and this has been my attitude for a number of years. But anxious as I am to wipe out the gulf that separates the two organizations, I will not agree to terms that will undermine and eventually destroy the great Farm Club machine which has been built up with such laborious effort during the last 15 years, and which merely stands on the threshold of its greatest usefulness."

Grant upfront that Hirth was biased. From Hirth's perspective, the most effective means of achieving one organization (as he stated at a meeting with Farm Bureau leaders) was to dissolve Farm Bureau and absorb its members into MFA, adding the Bureau's board members to MFA's board and "take these Farm Bureau members into the fold on terms of absolute equality with our own members."

In the background of this debate, Howard Cowden was hard at work as MFA's efficient secretary. He was busy cultivating board members, interacting with his staff of paid field men and streamlining MFA's business operations. He was enticing MFA board members to serve as paid field men while they were also serving on the board, tying them to him more closely still.

In 1926, he had two other items on his agenda: meeting MFA's fast-growing demand for motor oils and products and working behind the scenes with MFA's executive committee (members he had personally cultivated and placed) on his own proposal to merge MFA with Farm Bureau. According to Schuttler's "History of the Missouri Farm Bureau Federation," at a Feb. 9, 1927, meeting with Farm Bureau, "Cowden presented a plan for a new state organization, together with consolidation of the livestock marketing agencies in which the MFA and the MFBF were interested."

Cowden wanted to create the Missouri Agricultural Association, complete down to corporate structure with specified divisions and associated management positions: Produce, Grain, Livestock, Finance, Legislation, Purchasing, Legal, Accounting, Transportation, Insurance, Organization, Field Service and Education (history's cruel irony is that MFA eventually wound up with these same divisions). Enacting this would require two bold steps: first marginalizing and then defeating William Hirth.

Hirth, upon seeing the document for the first time much later, would envision the skeleton of a political machine run by the board's executive committee, much like the one employed by Kansas City's political boss Tom Pendergast, whose machine Hirth fought against mightily in the political arena.

Farm Bureau executives thought they were dealing in good faith with MFA as an organization. As it turned out, they were dealing with a faction that was actively trying to wrest control of the cooperative from Hirth.

All the while, Hirth, as chairman of the Corn Belt Committee, was in Washington leading the charge on the McNary-Haugen bill. He had left Cowden in charge of MFA activities, but remained unaware of the extent and nature of Cowden's plans. The renegade MFA bunch was about to learn a valuable lesson about poking the "Lone Grizzly," as Hirth had described himself in a letter to a friend.

Payback

Howard Cowden worked through Standard Oil to supply farm club orders for fuels and lubricants. Back in 1922 he had visited a Minnesota cooperative supplying those products to farmers. He'd been thinking about the process since. Cowden specifically focused on structuring and implementing a consumer cooperative that centered exclusively on the fuel products line.

For several years, many of the exchanges had been selling petroleum. First Hirth and then Cowden had pooled orders, negotiated volume discounts and signed a contract to supply those orders with Standard Oil Company. Cowden studied the situation extensively and soon began working out how to design his own cooperative that dealt exclusively with these products and supplied them to other cooperatives—a federated structure. MFA's contract with Standard Oil was to expire July 15, 1927.

Tensions between Hirth and Cowden had been escalating for three years. Twice before at earlier conventions, Hirth had proposed bylaw amendments requiring the Old Guard be responsible for signing farmers to the contract. Specifically, at a full board meeting in 1926, the board passed a resolution prohibiting board members from becoming field men. The resolution passed 17 to 11. Cowden, probably with the sanction of the executive committee, ignored the resolution. Board members stayed on his payroll.

Hirth knew he and Cowden were working at cross purposes. More to the point, he suspected Cowden of trying to control the board and oust

Hirth. Hirth wasn't wrong. Twice in the previous two years, members of the board's executive committee had questioned the practice of using The Missouri Farmer (a publication controlled by Hirth) as the official MFA communications vehicle.

To counter that question, Hirth twice proposed to the full board that The Missouri Farmer be withdrawn as the official organ. The full board voted against the proposal. Additionally, on Oct. 14, 1925, Hirth officially submitted his resignation from the board "finding myself out of harmony with certain policies." The full board objected his resignation and asked that he withdraw it, which Hirth did. But his objections were now out in the open.

As he reminisced in The Missouri Farmer shortly before his death in 1940, he and Cowden for the first several years had "worked together in harmony, but once he got in charge of the fieldmen he gradually began to assume increasing authority, and since their jobs were dependent upon him, they soon began to carry out his schemes and to end in ignoring my wishes. In order to increase his hold on the directors of the State Association, he placed a number of them on the field force payroll."

Adding fuel to the fire, Hirth soon learned of Cowden's proposal to restructure MFA and merge it with Farm Bureau. In the midst of the confrontation, in March of 1927 Cowden submitted his resignation to become effective at MFA's annual convention in Sedalia beginning Aug. 29, 1927. He still had the strong backing of the executive committee, members of which were firmly anti-Hirth.

The board, led by the executive committee, ignored Hirth's offered advice to hire R.J. Rosier (who would become one of MFA's most business-savvy executives) to replace Cowden and instead hired Frank Scott—a member of the board of Producers Produce Company of Chillicothe and a man openly critical of Hirth. Still in good standing with the executive committee and the president of the board, Cowden moved quickly before his resignation was in effect and in an executive committee meeting a month before the 1927 annual meeting proposed that he be allowed to fulfill the terms of Standard Oil's contract. The committee agreed.

MFA's 1927 annual meeting showed this fractured state of affairs. For the first time, the presidential candidate (the Cowden-backed T.H. DeWitt) was challenged by a Hirth supporter who narrowly lost. But the vice presidential candidate (J. Wiley Atkins of Lebanon, a Hirth supporter) won, leaving a divided leadership.

When Cowden officially formed Cowden Oil Company on Jan. 27, 1928, the oil company's board included some of those very same directors who voted to accept the contract, as well as DeWitt, now president of MFA and also a field man who had reported to Cowden.

Hirth assumed full attack mode. He wrote to Cowden and asked the following questions: "First, do you not think that our Association could buy oil just as cheaply as you can? Second, if so, how can your contract be justified? Third, why was not the ratification of your contract put up to the full Board in St. Louis? Fourth, who wrote your contract?"

Still on fire, Hirth announced the 1928 MFA annual convention would decide the future leadership of MFA, whether it be led by Cowden or Hirth. And decided it was. Just prior to the convention on Aug. 15, 1928, MFA President DeWitt mailed out a proposal (designed by

This feeder was donated by Gary Heldt of Cooperative Association #130 of Hermann and Rhineland.

Cowden) to consolidate MFA with Missouri Farm Bureau Federation and the Missouri Farmers Union. The merger would bring both Farm Bureau's and Farmers Union's livestock associations into the fold as well. "You will note that the principle [error in original] change, affecting the M.F.A. is in the name of the organization [Missouri Agricultural Association]. While we regret that the Committee deemed it necessary to make this change, we feel that is a secondary matter when it is realized that the benefits of the new organization can be made more far reaching," stated DeWitt's letter, which had proposed bylaw amendments attached.

The letter closed with: "I hope you will study the enclosed report with an open mind and base your conclusions with the benefit of the greatest number of farmers in mind. This report will be submitted to the State Board on the eve of the Convention for their adoption or rejection. However, the final decision on this matter rests with the Convention."

For several weeks leading up to mailing the letter, DeWitt and other members of the executive committee had been holding meetings to inform members about the benefits of the merger. One of those letters was forwarded to Hirth by Mrs. J.W. Spicer of Marshall along with her personal commentary. She and her husband (he attended the meeting) were having none of it.

"Mr. Spicer said there was no enthusiasm manifested by the audience and that 'twas simply an attack upon you—preconvention log rolling," she wrote. "He [her husband] was utterly disgusted. … We are highly incensed over the lack of principle displayed by the State Board and hope when the 'house cleaning' is over each of those dirty rascals will have been fully exposed, and sent home disgraced forever in the eyes of the Missouri Farmers Association and I can't believe there are enough unprincipled people in the organization to retain any of them."

As noted above, each year at the annual convention, MFA members voted on a slate of MFA officers to lead the organization. Hirth, considering the officer positions a post of honor and recognition, had always demurred when asked to accept the presidency saying it should be held by the men who built MFA. This time he embraced the offer.

This one-of-a-kind pin was worn at the first convention. "Tired of being the goat" was a rallying cry for the farm clubs and farmers who were united in finding and providing a voice for agriculture.

▲ Annual conventions provided a venue to elect the officers of MFA and to pass resolutions for business and political actions. This was MFA's first convention and it was held Aug. 29, 1917, at the fairgrounds in Sedalia.

▼ Dan Gause of Dallas County, who was vice president of MFA, challenged MFA President Ned Ball of Montgomery County to see which section (south versus north) could sign up the most new MFA members by March 1. If the north won, the MFA convention would be held at the Sedalia fairgrounds. If the south won, the convention would be in Springfield. The south won and the convention was held in Springfield in 1921.

Convergence—Politics, Federal Farm Board, Cowden, Farm Bureau and Paybacks

MFA Stores

Chamois

Clarence

Clarksburg

Clinton

Cole Camp

The parking lot on the Sedalia fairgrounds outside the convention shows the difficulty of long-distance travel. Many drove early automobiles; others came by horse and buggy or by train.

Eight thousand people and the largest-ever representation of voting delegates (1,750 out of 1,800 authorized) converged on Sedalia for MFA's 1928 convention at the fairgrounds. Tents were erected as temporary homes with makeshift kitchens. According to coverage in The Missouri Farmer, "Others slept on cots located in various buildings about the grounds while hundreds of people took advantage of Sedalia's hospitality, whose citizens generously opened their homes to the visitors at rates that were most liberal."

Whenever Hirth's name was mentioned by a speaker "the roar of approval that went up literally shook the foundation of the Coliseum. …Each time Mr. Hirth appeared upon the platform, if by only to make a passing remark, the audience roared its approval. At the mere mention of his name they cheered and when nominated for the Presidency the entire audience came to its feet with a shout like the roar of Niagara and when elected straw hats soared high in the air while the Convention shouted itself hoarse. It was overwhelmingly a Hirth crowd and the opposition was like the fallen leaves of Autumn before a windstorm."

Hirth was elected president, J. Wiley Atkins was re-elected vice president and board candidates supporting Hirth were overwhelmingly elected to replace dissidents. Hirth proposed an amendment allowing delegates instead of board members to choose the executive committee, eliminating the potential of the political machine he feared. The rout was complete. What's more, Hirth received the support of every single delegate from Polk County—Cowden's home.

CONVERGENCE—POLITICS, FEDERAL FARM BOARD, COWDEN, FARM BUREAU AND PAYBACKS

Another forceful resolution easily passed condemning the actions of the prior board who had:

> …during the past year utterly ignored the will of the overwhelming majority of the farmers who compose the membership of this Association and used their efforts to block and oppose the will of the membership even to the extent of engaging in a personal statewide campaign of vituperation and abuse and whereas this convention has once again, in language and by action that no man can misunderstand, expressed its will and chosen its policies, therefore, the board members thus referred to who have not abided by the will of the members of this Association as here expressed have served their usefulness as board members and their resignation as board members is requested in order to promote the best interest of this great farm organization.

The new board, under Hirth's leadership, gave Secretary Frank Scott his walking papers—an action he fought and lost. They replaced him with R.J. Rosier. Rosier, whose father was a strong MFA member from southwest Missouri, had managed the MFA elevator in Adrian and had since been an MFA auditor for seven years. Cowden's oil contract with MFA hit the skids.

Cowden was down, but not out. He continued through constant effort to convince individual MFA exchanges to buy from his company, keeping some of those dissident MFA board members on his board. His brightest future, however, came from reworking his company into a cooperative called Union Oil Company. Cowden thought interregionally. Eventually, Union Oil Company became Farmland Industries with Cowden as its president. A Farmland-MFA rivalry was born. It would last until Farmland's bankruptcy in 2002.

Building MFA Oil Company

The heat was on. MFA either needed to form its own petroleum cooperative or acquiesce to Cowden's onslaught. Hirth knew the stakes and focused obsessively on creating just such a business. Creating an MFA oil company wasn't a new concept, having been discussed at MFA board meetings as early as 1921, but now it was imperative.

To illustrate those intentions, the Phelps-Maries County Farmers' Produce Exchange at St. James received an Oct. 28, 1928, letter from the Chicago Equity-Union Exchange endorsing Union Oil Company at Kansas City and asking the MFA exchange to sign on.

"It is the intention of this organization to become a service organization for local

"It was overwhelmingly a Hirth crowd and the opposition was like the fallen leaves of Autumn before a windstorm."

—**The Missouri Farmer**

cooperative oil companies. We are personally acquainted with Mr. Howard A. Cowden, who was for several years secretary of the Missouri Farmers' organization, and the formation of his company has grown out of his experience in cooperative marketing with that organization."

The MFA board at its Sept. 26, 1929, board meeting, realizing Cowden's growing threat, addressed the situation with this resolution they broadcast far and wide:

> Apparently some of our members in different parts of the State are under the impression that a certain oil company is connected with our Association, and in view of the fact that certain officers of this company were formerly officers of the Missouri Farmers' Association, this makes such confusion easily possible. In this connection we desire to say that the Association has its own M.F.A. Oil Company which has our fullest endorsement because by patronizing it all the profits in this business will go to our members, instead of being shared in by outside parties, and we therefore, trust that our members and local leaders will assist in the financing of M.F.A. Bulk Oil Stations throughout the State.

In January of 1929, Hirth was writing refineries for terms. Letters from MFA exchanges came pouring in. They wanted petroleum products now. William Bret, manager of the Producers' Exchange at Bonnots Mill, wrote Hirth Jan. 11, 1929, and explained his exchange's oil setup:

> We have two bulk storage tanks of 10,500 gallon capacity each, one for gasoline and one for kerosene. Very shortly we expect to put up another bulk tank to handle anti-knock gasoline. The new high compression motors require this kind of fuel. We have one and one-half ton International Truck on which we have mounted a 500 gal. three compartment tank. We also have motor oil carrying cans that we can load on our truck when necessary. …Incidentally, I might say that it is my personal opinion that the M.F.A. ought to get behind some kind of plan or system regarding the oil business.

Hirth and Rosier were moving quickly. As Hirth wrote to Oscar Royse of Cape Girardeau, who was secretary of the Farmers Co-op Association there, "I have your letter and in my opinion the M.F.A. will enter the oil business in the very near future, not only on lubricating oils, but on gasoline and kerosene, confining ourselves to points where we can install storage tanks and unload direct from tank cars into such storage tanks."

In turn, Hirth was pushing other exchanges to think about installing stations in their communities as reflected in his letter to Otto Linsenbardt, secretary of the Cole County Farmer Association: "For your confidential information, I think we will begin financing bulk oil and gasoline stations in the near future, and these stations will save the average farmer several times his annual dues, and why not put one of these stations in Jefferson City?"

Writing to C.E. Lane, manager of Producers Produce Company in Springfield, Hirth asked him to raise $10,000, ideally by finding nine farmers willing to loan $1,000 each at 8 percent interest (MFA would contribute the other share). And, by the way, "without a moment's delay."

Cowden's Union Oil Company was rumored to be exploring the Springfield market, and Hirth wanted a head start. "But again," he wrote Lane, "I repeat that not a moment is to be lost." He was writing roughly the same letter to individuals statewide.

Demand was building. Hirth had learned hard lessons in structuring existing MFA-affiliated operations. In structuring and staffing the new oil cooperative, Hirth was determined to do it right. He drew on his experience to build a more tightly structured business.

To start, the board's new executive committee, chaired by Hirth, hired A.D. Miller, an experienced sales manager of a petroleum company, to build the business. But he was hired 30 days at a time pending performance. By June of 1929, MFA Oil Company had been formed with Miller at the helm, reporting to the executive committee and Hirth. It was located at 5473 Delmar Boulevard in St. Louis. By early 1930, 24 bulk oil plants were financed, constructed and operating. The first official delivery, marking the official beginning of the company, would be from the new Wright City bulk plant to William Wisbrock on Oct. 26, 1929.

Location of bulk plants sometimes engendered hard feelings since ready access to products and local construction were economically important to communities at the start of the Depression. One such letter survives in Hirth's papers at the Western Historical Manuscript Collection. It is from Hugo Hasenjaeger of Treloar who lists his town as a strong farm club center.

Hirth's reply showed his emphasis on letting business dictate policy. "As to what you say of the loyalty of our members at Treloar, all these things are true and deeply appreciated by me. … And yet when we hold those at the head of the Oil Department responsible for the best results, I am sure you will realize that I cannot at the same time dictate to them in such a vital matter as the matter of sites. For instance, I thought that our Bulk Station in the North part of the County would be located at Warrenton rather than Wright City, but again I kept hands off." Sound business principle dictated that approach.

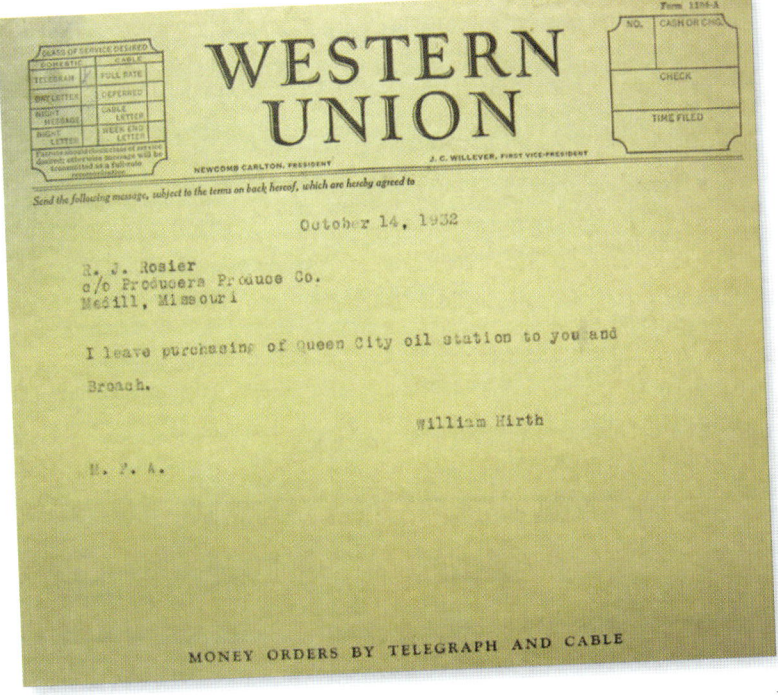

▲ William Hirth had learned valuable lessons in structuring different agencies around the state. By the time he helped create MFA Oil in 1929, he was determined to keep business decisions in the hands of those accountable for performance. (Western Historical Manuscript Collection)

▼ This 1929 Ford AA 1-ton features an aftermarket "Open Express" body. It is the first truck purchased exclusively for the new MFA Oil Company that, at the time, was headquartered in St. Louis. (Western Historical Manuscript Collection)

Simultaneously, A.D. Miller's business-like approach to running the new oil company is reflected in his Nov. 27, 1929, letter to MFA: "The more stations that we put in a producing position the larger our gallonage on both motor oils and greases as well as kerosene and gasoline the better the opportunity to carry our overhead and set aside a reserve."

But, Hirth, being Hirth, kept a close eye on expenses at this new venture, replying to Miller on Nov. 30 and scolding him for sending three letters in one day, each with its own 2-cent postage: "The three letters I received this morning could have come through for 2 cents postage as well as 6 cents. These things may look trivial, but in the course of a year they amount to considerable, and also if an office force is wasteful in one particular, it is very likely to be so in others."

A subsequent letter (Dec. 21, 1929) from Miller showed his dedication: "We are just as proud at the launching of this new ship on the sea of business as you are and we are guiding her to the best of our ability without regard to hours, putting forth the very best effort that each and all of us know how. I assure you will watch it with extreme care so that a year hence our success may be finally assured and that you may be wholly justified in anything you say about the M.F.A. Oil Department."

The new business showed immediate profitability with each new station operating at a profit. Within the first official month of business, MFA Oil Company moved more than a trainload of petroleum products.

The stark realities of the Depression would soon take hold, but so too did the new oil company. Sales volume increased each year for the next four decades. In 1938, Hirth made another wise move. He hired Raymond A. Young to be assistant manager of the oil company. Young would become its executive vice president, a position he held for 42 years. He would also become executive vice president and chief operating officer of MFA from 1968 to 1981.

Within its first official month of operation, MFA Oil moved more than a trainload of petroleum products.

MFA's new oil company competed aggressively with Howard Cowden's new oil company. By early 1930, 24 MFA bulk oil plants were up and running. The new company posted a profit each year for a decade.

> *"I assure you will watch it with extreme care so that a year hence our success may be finally assured and that you may be wholly justified in anything you say about the M.F.A. Oil Department."*
>
> —A.D. Miller

Proud Past, Bright Future: MFA Incorporated's First 100 Years

S.A. NORDYKE, PRES.
C.E. LANE, V. PRES.

W.B. ROGERS, SEC'Y
W.T. CRIGHTON, MGR.

PRODUCERS CREAMERY COMPANY

MANUFACTURERS OF

"LAND O' SMILES"

SWEET CREAM BUTTER
SWEET CREAM
AND
POWDERED SKIM MILK

555 W. PHELPS ST.

PHONE 3838

SPRINGFIELD, MO.

December 29, 1931

Mr. William Hirth,
The Missouri Farmer,
Columbia, Missouri.

★★Chapter 3★★
MFA in the 1930s

"Everybody's business is nobody's business"
William Hirth would live until 1940. His last 10 years would be productive—but more in the political realm than the business. He was at the height of his political and intellectual powers. Farm clubs at this point totaled almost 2,500, which operated 375 exchanges and elevators. MFA remained ascendant in 1930. Total U.S. farm population was 30.4 million. By 1933, in response to the Depression, the board of directors lowered annual MFA dues to $1 per year. And that dollar included a subscription to The Missouri Farmer. Despite the coming economic collapse, only three exchanges closed their doors during the 1930s. MFA would end the decade by celebrating its 25th anniversary in 1939.

To start the decade, MFA led efforts to enact a graduated income tax (which moved taxing emphasis from property to net profits) as well as legislation requiring the governor to prepare, in advance, a budget of appropriations for each state institution. In 1931, MFA publicly sponsored (and funded) the Committee on Taxation and Governmental Reform in Missouri. The committee was purposely bi-partisan and focused its deliberations on state, county and city government. One of the committee's first actions was to call the governor before the committee.

The results were to be dramatic and would result in a decrease in MFA's state political influence for a few years. Why? As Hirth was to point out in 1933, the measures enacted,

> stepped on the toes of a pile of politicians and job holders; under Constitutional Amendment No. 2, the legislature is limited to not employing over 150 clerks, stenographers, doorkeepers, etc., whereas previously it had employed between 800 and 900 and the resentment of the members against this was manifested by the employment of 68 additional 'janitors' by the Permanent Seat of Government Commission at $5 a day; also two years ago, we reduced the salaries of county officials and under the County Budget law county courts cannot spend more than their yearly revenue income, while clothing the State Auditor with the power to audit the books of county officials has turned up hundreds of thousands of dollars in shortages…

MFA Stores

Columbia

Columbia

Conway

Crane

Crocker

In addition, reported Hirth in The Missouri Farmer: "Through the school bill we have established a 20-cent levy for rural school purposes throughout the state, and infinitely more important, we have guaranteed eight months school, and placed a high school education within the reach of every farm boy and girl in Missouri."

More importantly from an MFA business perspective, though, was a 1930 convention resolution that gave administrative powers to the state association. That power extended to management of all elevators, exchanges, shipping associations, central plants and other MFA agencies. This major step toward a more centralized control paved the way for today's business model.

"Everybody's business is nobody's business," Hirth wrote in The Missouri Farmer. "The Association had much to learn from 'big business,'" he said. As Derr reported in "Missouri Farmers in Action," "The recently organized MFA Oil Company was an example of Association control, the by-laws delegating supervisory authority to the Association's board of directors. Hirth recommended that similar control be exercised over the central plants, milk plants, and creameries, as well as the smaller business groups."

Until that point, as noted before, MFA affiliates had been free to take the buffet option—choosing this practice, eschewing that. All MFA businesses now would be subject to a yearly audit, requested or not. Operating costs were all important. To Hirth's immense relief, by 1933 the members of Producers Produce Company of Chillicothe would vote to rejoin MFA, open its books to auditors and follow MFA-dictated financial practices.

"...we have guaranteed eight months school, and placed a high school education within the reach of every farm boy and girl in Missouri."

—William Hirth in The Missouri Farmer

"I want to strip to the waist for this fight"

William Hirth was a fiscal conservative. Despite the hundreds of millions of dollars in MFA business that Hirth oversaw, he cast a jaundiced eye on national economic conditions. He saw disaster, not simply for MFA but for the nation. Again, Hirth, being Hirth, turned concern into action.

Hirth was convinced the United States was facing an economic storm the likes of which the country had never seen. He feared revolution. After one McNary-Haugen debate, in which he was accused of painting too pessimistic a future, Hirth retorted his accuser would change his mind when countless farmers began to lose their farms and rural banks began to close their doors.

In June of 1929, Hirth wrote to a friend: "In my opinion, we are drifting into perilous water not only with reference to Agriculture, but in other premises—among other things I think we are approaching an unemployment period that will shake the Nation from center to circumference."

Rather than tone it down, Hirth became more outspoken and contemptuous of Hoover's involvement in the farm sector and the economy at large. He wrote articles, published in The New York Times, lambasting Hoover. He railed about Hoover's political handling of agriculture, charging waste of $200,000,000 of taxpayer money in large purchases of commodities. He called the Farm Bill a "shrieking farce" and the Farm Board's efforts "insane," noting most of those hired were being paid far more than their efforts in private industry would ever have allowed.

The wheat-marketing agency, he said, had hired more than 1,000 employees at exorbitant sums. The entire Farm Board, he charged, "provided the biggest buzzard's roost in the history of Governmental Departments." What's more, he said, the agency attempted "to put an iron collar around the organizations it could not control…first and last it is the most wasteful and demoralizing agency in the history of Government."

October 1929 would usher in the Great Depression. Black Thursday, Oct. 24, 1929, marked the official beginning of the Depression. It was followed by Black Friday, Black Monday and Black Tuesday, each bringing tremendous drops in stock values. The decline continued throughout 1930. By midsummer of 1930, share prices had lost almost 90 percent of their September value. Not until 1954 did stock regain 1929 levels.

Farm prices tanked. MFA, and the country, faced an economic onslaught. Hirth wrote:

> At this hour foreclosure and sheriff's sales of farm homes are the order of the day at every Court house in the land. The truth is that the vast majority of our 6,000,000 farmers are bankrupt, and yet only a few years ago farm loans were considered as safe as Government bonds. Nearly 10,000 rural banks have been compelled to close their doors, and these unprecedented failures graphically depict the almost complete collapse of rural life. But this is only half the story, for our towns and cities are filled with millions of idle workers who exist largely upon public charity, and whose children go to bed hungry at night, and this in a land which doesn't know what to do with its surplus food.

"In my opinion, we are drifting into perilous water not only with reference to Agriculture, but in other premises — among other things I think we are approaching an unemployment period that will shake the Nation from center to circumference."

—Letter from William Hirth to a friend, June 1929

Fearing an even greater U.S. economic morass, Hirth began a political collaboration that put him squarely in the midst of national power. One letter from Hirth dated Feb. 13, 1932, instructs a young governor that Hirth would be in Washington "the latter part of next week, and if you still desire to see me you may drop a wire to the Willard Hotel." The governor was Franklin D. Roosevelt.

No mention is made of a New York City meeting, but another Hirth letter (Feb. 26, 1932) to a friend in Canalou, Mo., states: "Incidentally, I took supper with Governor and Mrs. Roosevelt night before last, and I hope you are for him, for I think he is the best bet in sight, and his heart is absolutely right."

That supper meeting is referenced by a Hirth March 29 letter to C.M. Herndon of Lebanon: "I recently spent about four hours with him at the Executive Mansion in Albany, and took supper with him and his wife, and I did this to satisfy myself as to what Agriculture can expect of him—in my opinion he is the only man in sight who appreciates the importance of Agriculture to the Nation's prosperity, and I think he will make a great President."

Further in that letter Hirth reaffirms that "I am not a partisan," a refrain he would repeat again and again throughout his life to explain his political inconsistency. Hirth did support members of both parties but most frequently found himself campaigning for Democrats despite his fast friendship with Republican senators, representatives and governors. From Hirth's perspective, he was for agriculture, first and foremost. Party affiliations were superfluous. What mattered to Hirth were "fearless men." In the 1932 campaign, while campaigning for Roosevelt, Hirth would support a Republican for state senator in Missouri.

A March 5, 1932, letter from Hirth to Roosevelt references a mailed questionnaire "on your farm operations to which I trust you will

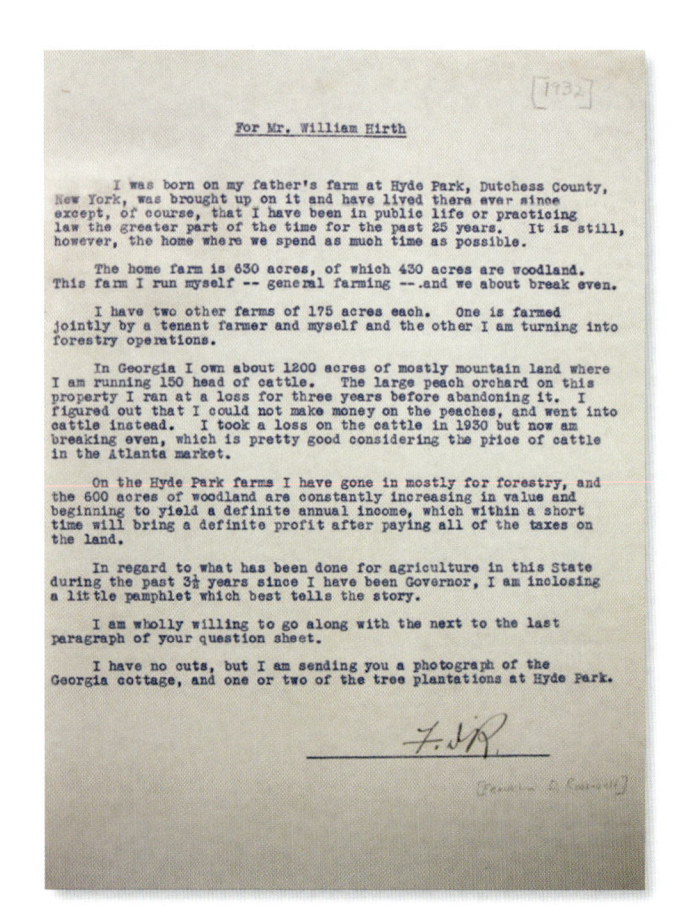

Proving that politicians of yesterday share similar genes with those of today, Roosevelt describes his Hyde Park estate as a home farm. There's a world of difference between a farm and an estate. (Western Historical Manuscript Collection)

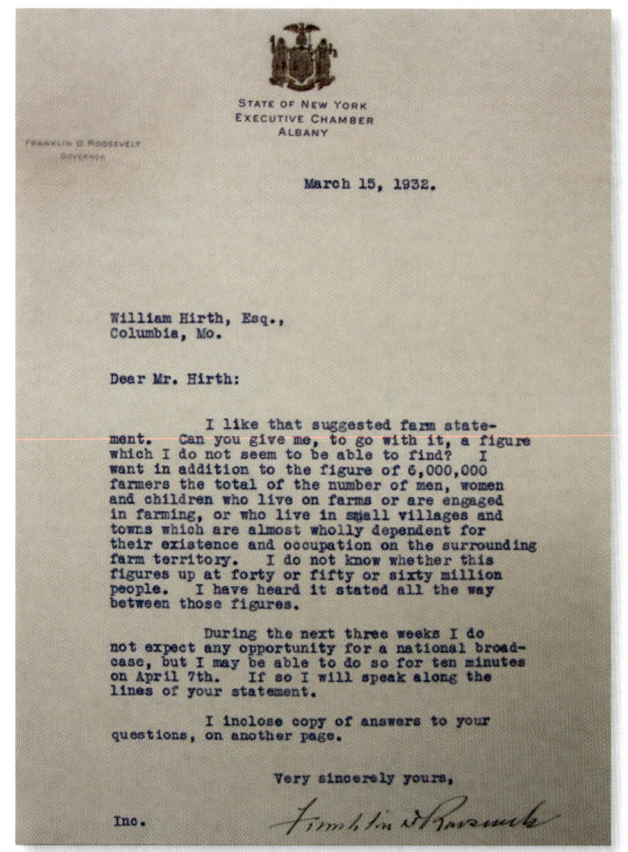

At the beginning of his presidential campaign, Roosevelt corresponded several times with Hirth and enlisted him to give radio speeches in support of Roosevelt around the Midwest. (Western Historical Manuscript Collection)

dictate an answer." The questionnaire survives in the Western Historical Manuscript Collection (as do most of these referenced Roosevelt letters). The letter accompanying it on letterhead of State of New York, Executive Chamber, Albany, is signed Governor Roosevelt and ends with "I inclose copy of answers to your questions, on another page."

The two would correspond throughout Roosevelt's presidential campaign. Many of those letters survive in the collection. Roosevelt wrote Hirth that "anything sent here to Albany will reach me promptly and have immediate attention." Hirth would become an important cog in Roosevelt's 1932 campaign, delivering radio addresses throughout the Midwest on request. His widely recognized contributions created a national buzz that Hirth would be the next U.S. Secretary of Agriculture.

Hirth's importance to the campaign is reflected in a letter written by Roosevelt to Ewing Y. Mitchell in Springfield. Mitchell and Hirth were friends:

"I hear from many quarters most enthusiastically about Mr. Hirth and his activities," Roosevelt wrote. "I have been in direct communication with him and, if you should see or write him in the near future, I wish that you would tell him that almost every letter brings me an account of his friendly activity on my behalf and that I am entirely aware of how much I owe to his untiring efforts."

Hirth submitted suggestions for Roosevelt's agricultural platform and ridiculed Roosevelt campaign officials who advised the platform not exceed 1,000 words. Hirth wanted more beef, as he suggested in June in a letter to Roosevelt: "…if we put out a platform with only 1,000 words or so, I think it would become a matter of jest—in this perilous hour the Nation's ills cannot be adequately discussed thus briefly."

He would be more forceful with his friend U.S. Representative Clarence Cannon: "…a platform of this brevity would be nothing less than idiotic…"

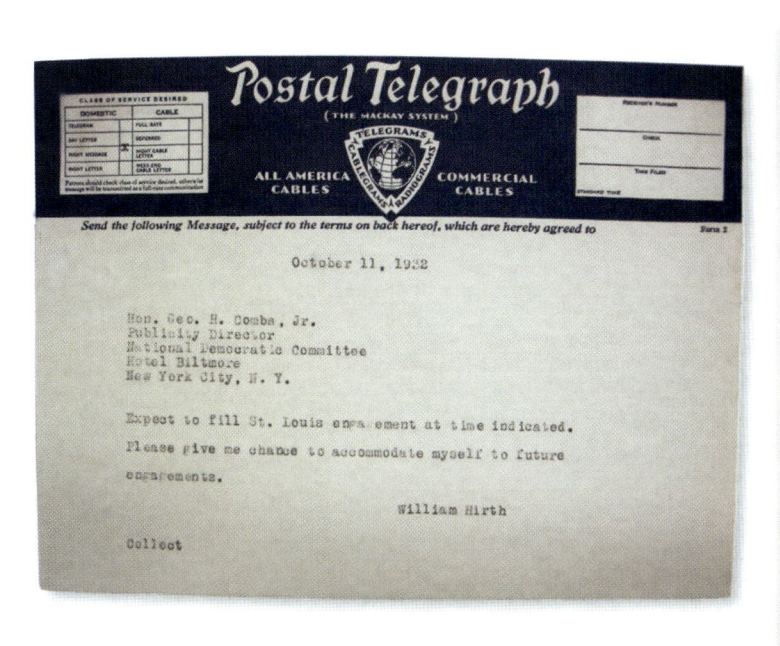

Hirth was willing to speak on the candidate's behalf, but he would not tolerate last-minute requests. His time was valuable. Notice that he sent the telegram collect. (Western Historical Manuscript Collection)

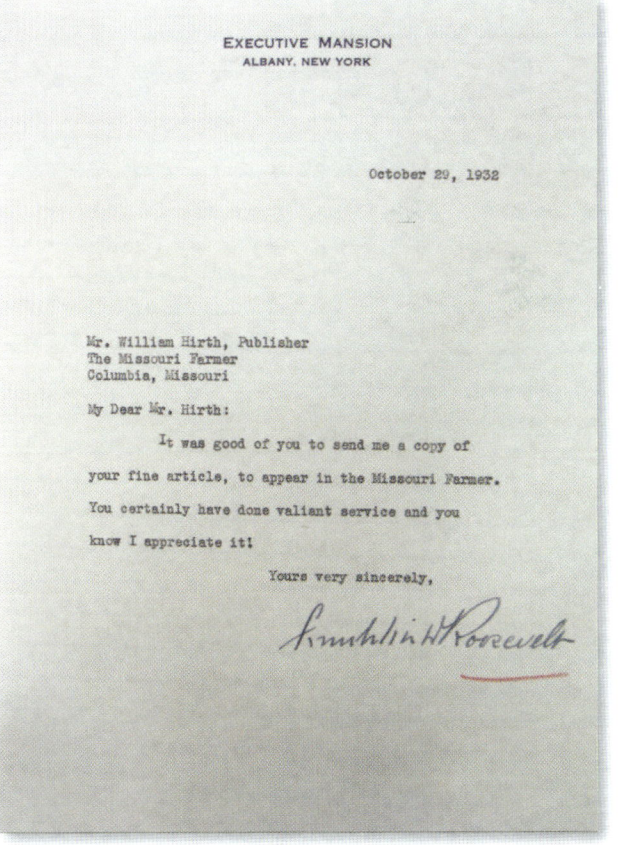

Hirth spent an evening in the New York Governor's Mansion with the future president and his wife Eleanor. He came away impressed with the governor's perspective on the farm problem. (Western Historical Manuscript Collection)

"...it will stiffen his backbone a lot... ."

—William Hirth to J. Frank Grimes
Referring to President Franklin D. Roosevelt

He would eventually reconcile himself with the platform but he could not see it as other than "bunglesome."

Underlying Hirth's support, however, was an uncertainty Roosevelt would perform once elected. Hirth urged his friend J. Frank Grimes, president of the Independent Grocers Alliance, to send Roosevelt Grimes's recent speech. "…it will stiffen his backbone a lot… ."

By July of 1932, Hirth was even more worried, writing a two-page detailed letter to Roosevelt after reading reports in the Kansas City Times of the candidate's preparation to release a statement on agriculture.

Hirth itemized his expenses and mailed them to Roosevelt's campaign headquarters. "I hope," he wrote in the accompanying letter, "that these addresses were worth the money. I have gotten a lot of letters from leading Democrats and also farmers. And I was especially interested in the latter . . ." (Western Historical Manuscript Collection)

How much of the article reflected facts and how much reflected

> the reporter's imagination, this I do not know, but since you have not as yet burned any bridges behind you, I am wondering whether I may not offer you a bit of unsolicited advice in these premises? The fact that I have built up here in Missouri the most powerful farm cooperative that can be found in any state in the Union, in fact one of the three largest active cooperatives in the Nation; and the fact that I have been the Chairman since its inception of the Corn Belt committee which bore the brunt of the McNary-Haugen battles, these things, I hope, make me more than passingly familiar with the so-called farm problem.

Hirth advised caution and enclosed his own farm statement, especially since Roosevelt was stuck with a farm platform that was, in Hirth's view, neither "intelligent [n]or constructive." "I was on the ground [at the writing], and shared the agony of the gods on this score, and therefore you will very largely be compelled to be your own farm plank…I know you will overlook the apparent presumption that is involved."

His unease continued to build. Writing to an official at the National Democratic Headquarters, Hirth parsed Roosevelt's recent comments on agriculture and warned the party to advise him.

> …I think the Governor's suggestion that 'no immediate prosperity to the Agricultural population of all parts of the United States' can be hoped for was unfortunate to say the least…I imagine that many farmers heard or read the above suggestion with a sinking heart…And now I come to the most serious 'blow hole' of all, as I see it, in the Governor's Topeka address, and I now refer to his suggestion that any plan that will make the tariff effective in our home markets 'must not be coercive,' and that 'the

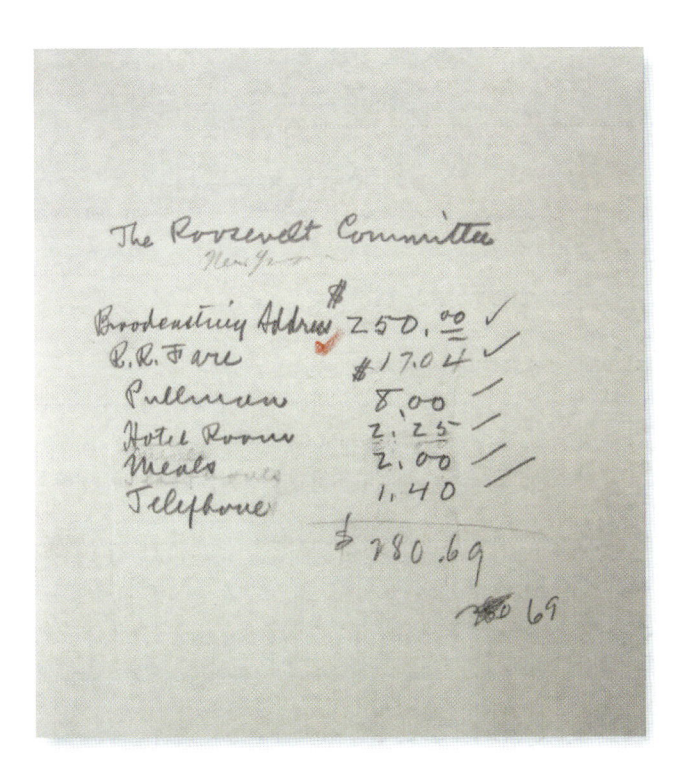

Hirth jotted down his expenses, starting with the radio fee. The station would not let him on the air without payment. Further letters testify to his displeasure. This was the Depression. Money was scarce. (Western Historical Manuscript Collection)

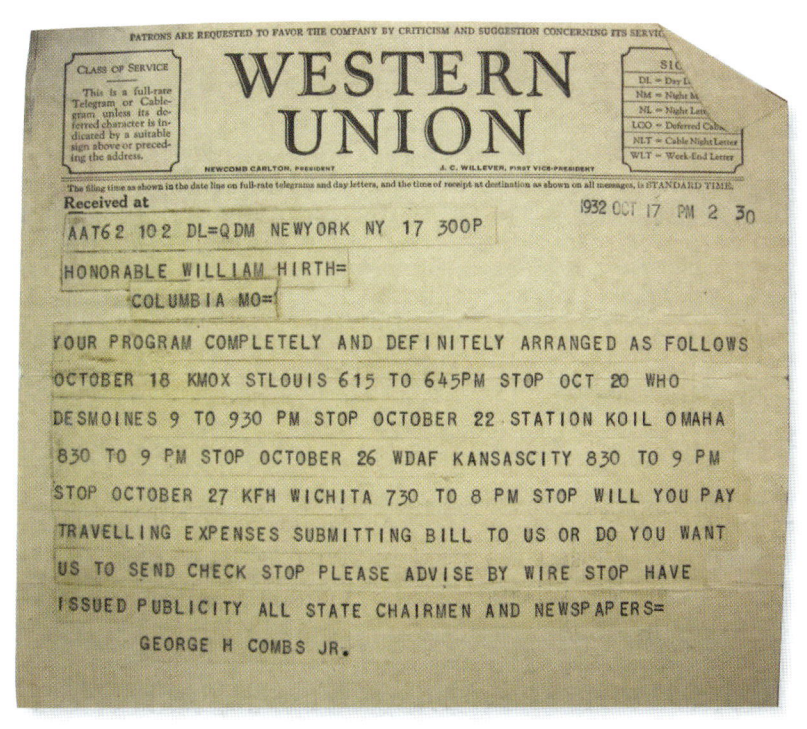

Hirth fulfilled all of the speaking engagements and won wide acclaim for his persuasive broadcasts. He was, however, miffed at having to pay for the Iowa broadcast out of his own pocket. (Western Historical Manuscript Collection)

individual producers should at all times have the opportunity of non-participation if he so desires,'…I am calling your attention to this matter because either I need to be put in an observation ward myself, or the Governor needs to trim his sails in these premises…"

Even on the day of the election, Hirth would continue to press Roosevelt to hold firm, writing, "And now I want to say that when again and again during the campaign you have stated that you did not believe that we can hope for a return of prosperity until the buying and debt paying power of our farmers and the millions who dwell in our thousands of rural towns and villages is restored, I have agreed with you unqualifiedly, and the immediate purpose of this letter is to express the hope that you will not permit yourself to be turned aside from this conviction, and that again you will not adopt a middle course…"

As to the secretary of agriculture position, Hirth would at first be coy, but then, after deliberation, he rebuffed all suggestions. "I am not hankering after any kind of a political office myself. I am for Roosevelt merely because I think he is the best man in sight," Hirth wrote a Missouri senator. Representative Cannon wrote Hirth Nov. 7, 1932, "Am very anxious to see you in Roosevelt's cabinet as Secretary of Agriculture." Cannon mentioned others nationally who were aligning themselves behind Hirth. Letters of support poured in.

Hirth abjured. He wrote straight to Roosevelt Nov. 14, 1932, and said: "I think a President should be permitted to select his Cabinet free from outside wire pulling, and furthermore when I said to you last winter that I would never be an applicant for a federal job if you became President, I meant exactly what I said." Historians and academics have ridiculed Hirth's rejection of a position he had not yet been offered.

Still, as the topic became more and more public, his denials would get more forceful.

Writing to a friend in Maryville, Hirth explained his rationale. "I have your several letters, and this is to say that I am asking my friends not to press me for Secretary of Agriculture, or any other Federal job—I want to help out as much as I can in a sound solution of the farm problem, and I will be greatly handicapped in doing this if I am placed in the attitude of being a mendicant for political favors."

Proud Past, Bright Future: MFA Incorporated's First 100 Years

"…I have spent the best years of my life in fighting the farmer's battles as effectively as I knew how, and now that there is a chance to 'cash in' I want to strip to the waist for this fight."

—**William Hirth**

To the secretary of a farm club whose members wanted to push his candidacy, he wrote: "…while I deeply appreciate this attitude on the part of your members I hope you will not pursue this matter further…I have spent the best years of my life in fighting the farmer's battles as effectively as I knew how, and now that there is a chance to 'cash in' I want to strip to the waist for this fight."

Professors and hairsplitters

Personally, Hirth was fuming. "Apparently he [Roosevelt] is surrounding himself with Professors and hairsplitters who are long on theories, but who don't know what it means to have a bloody head as I have from daily contact with closed banks, and the marketing of millions of dollars' worth of farm commodities," as he wrote to his friend Col. James Thompson, publisher of a New Orleans newspaper.

It was not his first nor last lament about professors and hairsplitters, a group for whom Hirth reserved contempt. When Henry Wallace was mentioned for secretary of agriculture, Hirth replied to a friend: "…[M]ore important than anything else Mr. Wallace is a hair-splitter, and I think such a man at the head of Agriculture at this time would be nothing short of tragic. Personally Wallace and I are good friends, but in my opinion we need a Secretary this time who has hair on his breast."

By 1934, Hirth was scathing in his denunciation of Secretary Wallace and Roosevelt's policies to date, writing directly to Roosevelt, explaining farmers were asking,

> How much longer can Uncle Sam keep on taking billions out of the Treasury to 'prime the pump?'…and thus it is not surprising that the average citizen is becoming alarmed when it is proposed to appropriate new billions of dollars for the purpose, and that he is wondering more and more how long Uncle Sam will be able to stand it without straining his credit to the danger point. …In my humble opinion you can no more hope to restore the Country to normal business conditions by mere artificial stimulation than you can make water run up hill. …As I see it we should stop creating fictitious purchasing power by piling up tax burdens for the future…

To make certain he was not accused of criticizing Wallace behind his back, Hirth sent a copy of the letter to Wallace as well.

Hirth saw many New Deal programs as "cumbersome, complicated, unworkable, un-American and unconstitutional." He raged that agriculture needed fewer "pointy-headed theorists" and more "double-fisted practical men" who had faith in old institutions and a full dose of farmer common sense. "…[I]f the plausible young theorists in Washington are right in their many new and strange ideas, then I am hopelessly ignorant in the above premises, and the thought that I should be regarded as a 'farm leader' is a joke."

By 1936, Hirth had had enough and publicly broke with Roosevelt over Roosevelt's attempt to pack the Supreme Court with six additional justices in order that he might pass New Deal programs the court had rejected as unconstitutional. Hirth wrote (in a news release scheduled for afternoon of March 17th),

While I am aware that it would have made little difference, I certainly would not have supported him for re-election had I been advised of what he had in mind with reference to the Court… Stripped of all specious pretexts and subterfuge, the President wants a Supreme Court that will o.k. the kind of legislation he desires, and if Congress grants him this unprecedented power, will not the Court cease to exist as a co-ordinate branch of the government?…[F]irst, that he desires a Court that will say that the Constitution is what he wants it to be, and second, that he has no patience with amending it through a 'solemn referendum' by the people, as was provided by the immortal framers of this historic document. And just as certain as Congress places this astounding power in his hands, the end of Constitutional government in the United States will be at hand…"

To cement his feelings concerning the New Dealers and their approach to problems facing agriculture and the nation, Hirth, shortly before his death, would write: "With all its faults, in my opinion, our country is the greatest in the world, and the chief hope of civilization, and, therefore, I have no patience with the young jackasses who want to 'make it over.'"

The major accomplishment of MFA as an organization throughout the decade was to hold and gradually build membership in an era of bankruptcy and ruin. Membership had grown to 26,659 farmers.

William Hirth would be elected MFA president for the final time during the 1940 annual convention, a gesture more tribute than expectant. Hirth's health had steadily deteriorated at the end of the decade. Because of his health, Hirth had missed the 1939 convention but mustered the strength to address the membership on the 25th anniversary via radio broadcast. His health worsened throughout the spring and summer of 1940. He died Oct. 24, 1940, at the age of 65. His death came one week and one day after groundbreaking for the new MFA building in Columbia.

He [Hirth] raged that agriculture needed fewer "pointy-headed theorists" and more "double-fisted practical men" who had faith in old institutions and a full dose of farmer common sense.

MFA Stores

Dixon

Downing

El Dorado

Eldon

Emma

A young Fred Heinkel

★★ Chapter 4 ★★
Passing the Torch

Out of the corn rows into the board room

Fred Victor Heinkel kept a sharp eye on the future. At the dawn of the 20th century, the young man farmed with his father in Franklin County. In 1917 rumors were swirling. Heinkel heard tales of a passionate individual proselytizing farm unity at speeches throughout the state. Young Heinkel was all ears. That individual, William Hirth, preached a revolutionary approach to organizing farmers—self-interest. Heinkel enjoyed farm life, but he was a realist, too. Making a living on a turn-of-the-century farm was a touch-and-go affair, so much so that the young farm boy constantly kept an eye out for money-making opportunities to supplement meager income.

William Hirth's tireless evangelizing included pamphlets. He printed and distributed them by the thousands. Heinkel latched onto one early, read it, kept it and chewed on the idea. He made a mental note to attend one of Hirth's frequent speaking tours. In April of 1917, Heinkel cemented his future. Hirth would be speaking in Jeffriesburg, a few miles distant. Heinkel, farming in partnership with his father William, urged his father to attend as well.

Fred Heinkel and his father signed this MFA Producers Contract while farming together in Franklin County. (Western Historical Manuscript Collection)

C. M. STILES, PRESIDENT T. H. DeWITT, VICE-PRES.
ALDRICH, MO. GREEN CITY, MO.

MISSOURI FARMERS' ASSOCIATION, INC.
HOWARD A. COWDEN, SECRETARY-TREASURER

COLUMBIA, MO. November 4, 1926

Mr. F. V. Heinkel
Robertsville, Mo.

Dear Sir:

 The election is over - <u>but not our work</u>, for regardless of what Congress does this winter, the farmers' problems will never be solved until powerful and efficient cooperative marketing associations are organized. The leading thinkers and economists of the whole Nation all agree on this. In this connection I am enclosing copies of statements made by two of the leading educators of the State. You will note Dr. Brooks, President of Missouri University has signed the Producers' Contract. We have not made Dr. Brooks' letter public nor have we announced before that he has signed. We are giving <u>you</u> this <u>advance information</u>. From time to time during the winter we expect to give you <u>advance information</u> about other important matters. We believe you can and will use it to good advantage. We are going to do this for the reason that the State Board is going to expect much more of you this winter than ever before. The Board at its recent meeting outlined the duties of Old Guards as follows: (1) Attend all meetings of his Club and take an <u>active</u> part; (2) Attend all County Conferences; (3) Help collect dues; (4) Serve on Speakers' Bureau; (5) Take an <u>active part</u> in the Contract Drive during November and December in cooperation with the County Board and County Secretary.

 In compliance with the action of the State Board a County Conference will, no doubt, be called immediately in your county. I want you to be sure to attend this meeting and help make plans for the signing of Contracts during November and December. This is very important for the reason that I want to be in position when the State Board meets in December to recommend that the Contracts in several counties be <u>thrown into effect</u>! We are not making a general announcement of this matter but I want you to be fully informed. Wouldn't it be fine to be able to throw several thousand Contracts into effect January 1, 1927? <u>If we are to do so we must all do everything within our power to sign every Contract possible during the next</u> eight weeks! Isn't this an opportunity you have long looked for - for us to get to the place where just a little more hard work on the part of each of us will put us into position to throw some of the Contracts into effect! <u>Then let's make the most of the opportunity!</u> Already the Old Guards are on the firing line in many communities and just think what an army we will have when the Old Guards from one end of the state to the other swing into action. The battle line in your county must not waver, the Old Guards of other counties are going to depend upon you to "go over the top" with them. Are you ready?

 I am enclosing a return envelope. I want you to use it in writing me about how much help you can give before the State Board meets the last of December.

 Your friend,

 Howard A. Cowden
HAC:RK Secretary

P. S. -- We are having a copy of Circular No. 150 issued by the College of Agriculture sent you under separate cover. <u>Watch for it</u>! <u>We suggest you read and discuss these statements at meetings you attend</u>.

Addressed to a young Fred Heinkel and signed by Howard Cowden, this letter was in Heinkel's personal folder at MFA. It shows another aspect of his activity during the formation of MFA.

At the meeting, Hirth extolled the benefits of cooperative marketing. Impressed and moved by Hirth's words, both Heinkels joined the farm club movement on the spot. Heinkel would never forget that first meeting. Thrilling, he would recall in interviews. "You young fellows are going to have to fight this battle differently than your fathers fought it," Hirth intoned to the group. Heinkel took that quote to heart, mentioning it time and again in interviews for almost 40 years.

Heinkel, just three years prior to meeting Hirth, had completed sixth grade (not much else was available to rural students who couldn't reach town schools) and passed a teacher's examination. But at age 16, he was ineligible to teach until 18. No matter. Teaching was a paltry living as well. In the interim, Heinkel redoubled his efforts at farming—and anything else that added to his income.

Fred Heinkel had three dairy cows, three stock cattle, 80 hens, 10 hogs and no sheep. He sold dairy and poultry products. His portion of the farm comprised 50 acres, 35 of them tillable. Years later, childhood friend Howard Kommer would write Heinkel a letter reminiscing about planting corn beside the creek. "I can still 'see' you plowing with the Fordson tractor and the 'new' type plow," Kommer wrote. "That farm was always a very bright spot in the farming community and very productive."

> *"You young fellows are going to have to fight this battle differently than your fathers fought it."*
> —**William Hirth**

Fred Heinkel had three dairy cows; his father had several more. The young Heinkel signed up as a member of the MFA Creamery located in St. Louis and invested $100 at 8 percent. (Western Historical Manuscript Collection)

Proud Past, Bright Future: MFA Incorporated's First 100 Years | 71

After hearing Hirth, Heinkel would immediately join the Pleasant Valley Farm Club. He traded at Catawissa. Over the course of the next few years, now aflame himself, Heinkel would first be elected secretary-treasurer of that farm club. He would then serve in the same role with the newly formed MFA livestock shipping association in nearby Robertsville and proceed to help organize the MFA Exchange at Catawissa in 1926. By 1931, he was elected president of the Franklin County Farmers Association, a position to which he was re-elected again and again.

The first recorded mention of Fred Heinkel in Hirth's papers at the Western Historical Manuscript Collection at the University of Missouri (from which the letters above are taken as well as Heinkel's treasured Hirth pamphlet) is a 1929 letter to William Ogles of Catawissa to whom Heinkel had sold a subscription to The Missouri Farmer. He was also signing farmers to MFA's producers contract. Later that same year, Heinkel took out an automobile insurance policy from Hirth's salesman on a "new Tudor Model 'A' Ford." The year's premium for fire, theft, personal injury, property damage and deductible collision? $36.27. As previously mentioned, as a sideline Hirth also sold insurance to supplement his income.

In that same letter (in a display of his industriousness), Heinkel asked for "the privilege of selling some of your insurance in Franklin and St. Louis counties." Not the exclusive right, Heinkel maintained, "just for the privilege of doing what business I may, in a spare time way, or as I meet up with people on other business." He also bought a couple of gilts from Hirth's purebred livestock farm.

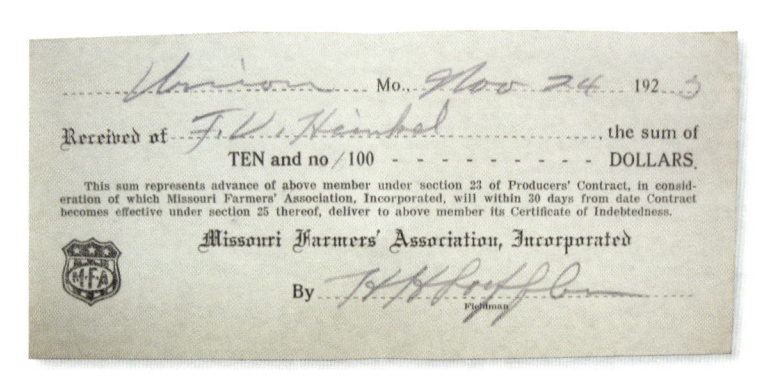

(Western Historical Manuscript Collection)

The "field work" notation on this check isn't what you might think. Field work meant functioning in furtherance of MFA initiatives by signing up members and doing other work related to MFA activities. (Western Historical Manuscript Collection)

▲ Hirth died one week and one day after groundbreaking for the new MFA home office building in Columbia. Heinkel consolidated MFA's central functions in the new office. Different departments had previously been spread over several different Columbia office buildings.

▼ This plaque was on the corner of MFA's original home office in Columbia, Mo.

By 1933, Hirth (through field man C.L. Cuno) was pressing others in MFA to convince the young Heinkel to serve as a county judge because the elected judge had died immediately after being elected. Heinkel, at the time, was a Republican in a county with 2,000 more Republicans than Democrats. When the nomination fell through (even though as local MFA leaders would write "he is beyond question the most able man in the county"), Heinkel did run for state representative. And lost. But his stock rose in MFA.

Hirth smelled talent and took to sending Heinkel position papers and asking for the young farmer's opinion. "I inclose for your confidential information copy of a letter which I have written to Secretary Wallace [U.S. Secretary of Agriculture], and concerning which I will be glad to have your frank opinion," wrote Hirth in one 1934 letter. Heinkel's opinions must have been sound.

By 1936, Heinkel was on the statewide ballot to serve as vice president with MFA President Hirth. Hirth took the 39-year-old farmer under his wing, introducing him to powerful people, explaining his objectives and outlining his plans for the future of MFA. Four years later, when Hirth died, Heinkel would, in his words, "Walk out of the corn rows and into the board room."

He walked into a mess. When Hirth died October 24, 1940, storm clouds of a world war obscured the horizon. Japan eyed Pearl Harbor and schemed. The nation was jittery with foreboding. And before the August 1941 convention, Heinkel, himself, would face a sneak attack on his newfound presidency.

PROUD PAST, BRIGHT FUTURE: MFA INCORPORATED'S FIRST 100 YEARS | 73

(Western Historical Manuscript Collection)

The Missouri Farmer speaks for MFA

Fred Heinkel lost no time in reassuring the membership that he would continue Hirth's policies, would deviate from them only when conditions so mandated. Not hesitant, though, in his new role, Heinkel announced five initiatives of his own: 1) that the managers and boards of the major agencies would be consulted more frequently; 2) that more meetings of local boards and employees would occur; 3) that a state legislative committee would be formed; 4) that legislative influence would be expanded by closer cooperation with other organized groups whenever the state legislature was in session; 5) that schools for MFA employees would be established to increase efficiency of all MFA agencies and to train men for the position of local managers.

The preponderance of these steps was political in nature. Only the last clearly focused on a business concern: training to promote effectiveness and efficiency. That emphasis underscored what would become Heinkel's major attribute: accomplishment in the political arena. The obverse would be true as well.

Make no mistake: Heinkel would become a business visionary. Under his leadership, MFA would build a business volume unequaled as well as a national power structure unique in U.S. agriculture. Under Heinkel's guidance, MFA would launch business ventures Hirth only dreamed of. Heinkel would strengthen the cooperative through a vastly diversified structure in all areas of farm production.

Just prior to Heinkel assuming the presidency, a committee composed of heads of major MFA agencies, at the urging of the board of directors, had met to consider a new approach to defining MFA's membership. As vice president, Heinkel participated—and led.

To this point, individuals paid yearly dues to MFA. Managers canvassed the countryside collecting dues, a time-consuming process. The new committee's recommendation, however, was an earned membership plan. This concept meant that when a farmer did $25 worth of business at an MFA agency, the farmer would automatically become an MFA member. The agency conducting the business would be responsible for paying half the farmer's dues to MFA (50 cents). The remainder was to be allocated between major agencies like MFA Milling, MFA Oil, MFA Grain and Feed, and the others. Once the plan was in effect in mid-1940, membership began to rise significantly. According to Ray Young, Heinkel, almost giddy with anticipation, was certain membership, in this arrangement, could rise to as high as 60,000.

Simultaneously, just before Hirth's death, MFA, under purview of both Hirth and Heinkel (Heinkel was on the initial committee), founded MFA Central Cooperative as a separate cooperative. Many exchanges, wobbly from the Depression's

▲ This one-of-a-kind blank certificate signed by Fred Heinkel is in his private folder at MFA. The training school was one of his initial accomplishments after taking the reins at MFA.

▲ This undated photograph shows a class of MFA employees who were enrolled in MFA's new training program. Heinkel wanted to increase efficiency across all agencies and saw training as a key first step.

PROUD PAST, BRIGHT FUTURE: MFA INCORPORATED'S FIRST 100 YEARS

weight, still floundered. MFA's purchasing department had been taking over the operations of those exchanges needing management. With the formation of Central Cooperative, the purchasing department moved the exchanges into one administrative body for management and control. That cooperative would become the backbone of MFA's retail distribution network of company-owned locations. By 1940, MFA Central Cooperative had five exchanges: Marshall, Versailles, Barnett, Eldon and Moberly.

But first things first. Heinkel had no sooner warmed the presidential chair than he and the executive committee were approached by a contingent of exchange managers demanding changes. They wanted the new membership plan scrapped, a plan of which Heinkel heartily approved and helped design. He would stand fast on this. Additionally, they wanted MFA to purchase and control The Missouri Farmer, the lifeblood of MFA. Whoever controls the magazine controls the cooperative, they opined correctly. Hirth had owned the publication lock, stock and printing press. Hirth's estate gave the publication as well as all other assets to his son, William Jr., who wanted to assume his father's role as editor.

Heinkel was a step ahead here. He and the board already knew they needed control of the magazine. He also knew negotiating with Hirth's widow and son would prove a ticklish affair. The acting editor of The Missouri Farmer was H.E. Klinefelter, a Franklin County boy himself. He and Fred Heinkel had worked together on the farm club movement. Klinefelter had a personal relationship with both Hirth and Heinkel that spanned decades. Klinefelter had become one of Hirth's first staff writers. With Hirth constantly involved with political and business matters, over the years he had turned over more and more of the editorial functions to Klinefelter. Klinefelter was torn between the two sides in the

When William Hirth died, his estate transferred ownership of The Missouri Farmer to his wife and son. Knowing the magazine was the voice of MFA, Heinkel immediately undertook negotiations with William Hirth Jr. and eventually purchased the magazine.

76 | Passing the Torch

MFA's purchase of Square Deal Stock Tonic was part of the negotiations with Hirth's family. The tonic's formula included iron sulfate, copper sulfate, napthalin and charcoal.

negotiations but knew MFA, which paid his salary, should control the magazine—and thus its destiny.

When young Hirth threatened to keep publishing the magazine independently, Heinkel and the board told him to go ahead, but since the MFA membership list was not his property but theirs, they would withhold it. With no one to send his magazine to, young Hirth finally agreed to sell his father's magazine to the cooperative for $25,000.

Heinkel's successful negotiations with the Hirth family brought The Missouri Farmer (and printing plant and Hirth's formula for his stock tonic, Square Deal) into the MFA fold for the first time. This strategic success helped cement the magazine as MFA's voice in agriculture.

Recognizing MFA's importance even with a new leader, U.S. Secretary of Agriculture Claude Wickard appointed Heinkel to the Agricultural Advisory Council of USDA, further elevating the young leader's stature.

MFA Stores

Enon

Fair Grove

Florence

Fordland

Forest Green

Internal strife is toxic

No sooner than Heinkel had solved the magazine issue and rebuffed changes to the new membership plan than another challenge confronted him. Opposition to his presidency arose from a group of delegates and managers dissatisfied with the manner in which the MFA Grain and Feed Company was organized. Their solution? Run a sympathetic man for president in opposition to Heinkel. Receiving word of the plan, Heinkel wrote a detailed letter to MFA leaders, asking for organized action in his behalf:

> As president of the M.F.A., it becomes my duty to warn you of a dangerous situation which will confront the delegates to our 25th Annual Convention.
>
> I have just learned that a small group in the western part of the State have held a series of meetings during which they claim to have already decided who the next president of the Missouri Farmers' Association shall be. It is my understanding that this group is composed most of managers, and that they claim to have rounded up a sizeable delegation which will vote as the group directs. …
>
> The point to bear in mind is that if this group has the power to elect a president, it also will have the power to capture the entire Association by electing an Executive Committee and a board of directors to its own liking. Remember—there are 19 directors to be elected!
>
> If a time has come when a small group can meet prior to a Convention, name a president, Executive Committee and directors and proceed to elect them, then we have reached a point where our conventions will become a farce, mere window dressing, since the membership at large will have little to say about the affairs and policies of the M.F.A.
>
> Some of this group are doubtless well meaning, but apparently there are a few of them who will not be satisfied unless they are in a position to run the entire Association! …
>
> Mr. Hirth often warned against such tactics as this clique reputedly is employing. He pointed out that if the M.F.A. ever falls it will be the result of internal strife occasioned by selfish or misguided people.

Heinkel's appeal and his campaigning were successful. He tallied a large contingent of delegates, 1,047 to his opponent's 457. But consider: After Heinkel's successful election at the 1941 convention, no punishment was exacted; no one was forced out. In fact, the opposition after defeat, declared themselves placated, their grievance acknowledged by the membership's vote. Their opinion had been duly aired and considered, they maintained. Graciously, they offered an amendment making Heinkel's election unanimous. And so it stood.

Over his years as president, Heinkel would respond to the world around him on his own terms. As a confidante of Hirth while Hirth's vice president, Heinkel was intimate with and awed by Hirth's national prowess. Where Hirth actively disliked the political side, he saw its necessity to achieve goals. Heinkel, however, was a natural on a political stage. Where Hirth was bombastic, Heinkel was quiet but tremendously effective.

Having watched the 1928 spectacle from the sidelines, Heinkel was impressed by Hirth's ferocity and head-on approach to the political fight. Heinkel, facing a familiar membership challenge that confronted Hirth, lobbied the membership but did not name, confront or challenge his opponents.

Heinkel was his own man, a man whose personality and approach stood in stark contrast to Hirth's. Hirth would dominate other strong leaders and bend them to his will. Heinkel was more adept at the softer art of political persuasion. He would perfect this craft over the years, lobbying MFA board directors personally, showing them his national contacts, pointing out his accomplishments and convincing them to do his bidding while serving on boards of MFA businesses. It was a practice that led to many of Heinkel's accomplishments, but a practice that also courted disaster. His effectiveness at influencing board members led him to weave a complex web of director placement in all cooperatives. The system worked well when directors could be counted on to follow his lead.

MFA had begun in 1914 with two philosophies, sometimes complementary, sometimes conflicting: improving farm life and improving the economic position of farmers and ranchers. To be successful, in the first part of the last century, the cooperative had to fill both roles, a practice that contained both great risks and great rewards. Very few individuals could be adept at both. William Hirth and Fred Heinkel came close to that ideal.

When Fred Heinkel walked out of the corn rows and into the board room in 1940, MFA was accelerating. MFA had operations, in today's lingo, from gate to plate. Producers Creamery, Producers Grocery Company and MFA Milling Company had a combined annual business of nearly $10 million and net earnings approaching half a million. The 11-year-old MFA Oil Company, begun and ably structured by Hirth and expertly organized and run by Ray Young, was perfectly positioned to serve the economic boom following World War II when returning farm boys brought machinery knowledge home and dreamed of gasoline-fueled farm expansion. Heinkel lost no time in tying MFA enterprises to the surge of post-war demand and production.

Heinkel, a visionary, dreamed great dreams but anchored them in the practical world of agriculture. As you'll see, he was both a highly competent politician and a tireless business entrepreneur. He was not, however, a businessman obsessed with solvency ratios, balance sheets, net worth or working capital. He was locked into a building mode, not a structural mode. While he knew balance-sheet terms and their significance, he lacked the business manager's passionate focus on detail. It would remain a flaw.

His genius in the political process and his lack of passion for business metrics and structure underscored a basic MFA contradiction: MFA's tremendous successes that built multimillion-dollar companies from scratch and MFA's failures in soundly structuring and tethering businesses to make one, solidified business enterprise.

Ray Young proved a steady hand on the tiller of MFA Oil Company. Young's hiring had been approved by Hirth, and Young would serve with Fred Heinkel for the next 39 years.

MFA had operations, in today's lingo, from gate to plate.

PROUD PAST, BRIGHT FUTURE: MFA INCORPORATED'S FIRST 100 YEARS | 79

Proud Past,

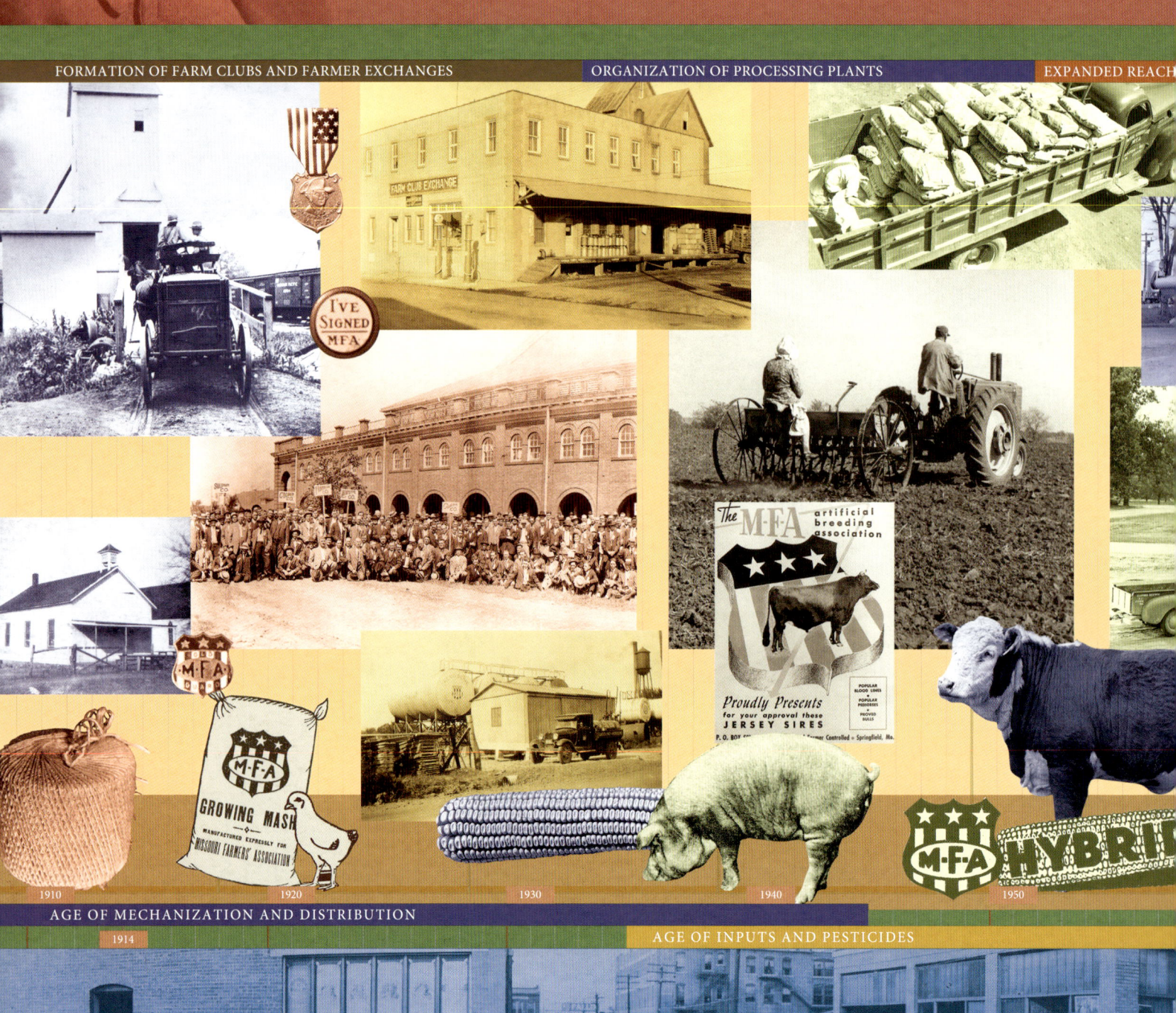

FORMATION OF FARM CLUBS AND FARMER EXCHANGES

ORGANIZATION OF PROCESSING PLANTS

EXPANDED REACH

1910 — 1920 — 1930 — 1940 — 1950

AGE OF MECHANIZATION AND DISTRIBUTION

1914

AGE OF INPUTS AND PESTICIDES

Bright Future

ERREGIONALS

FINANCIALLY SOUND ENTEPRISE THROUGH PROFESSIONAL MANAGEMENT

1970 1980 1990 2000

AGE OF TECHNOLOGY

AGE OF BIOTECHNOLOGY

At the dawn of World War II, Fred Heinkel led MFA's expansion in membership, businesses and political activities. During this decade, MFA would join the National Council of Farmer Cooperatives to provide a consistent voice in national policies.

★★ Chapter 5 ★★
Fred Heinkel Takes Charge

Explosive growth in the face of a world war

Heinkel, in conjunction with the University of Missouri's College of Agriculture, opened MFA's first employee training school in 1941. His intent? Teach employees the history of the organization, the extent of its business operations and, most importantly, the principles and philosophy behind the concept of cooperatives. The school would educate hundreds of employees during the next several decades and build a foundation for growth and employee talent.

But 1941 also brought total U.S. involvement in World War II. The world's existential struggle brought a whole new set of problems to agriculture. Farm boys flocked by the thousands to the armed forces at precisely the time food production needed to increase. Rationing of the everyday items like gasoline and tires brought a whole new reality.

By early 1942, Heinkel was forced to lobby Washington and the Office of Price Administration to complain of federal rationing rules. Heinkel, appointed by state politicians as the Missouri chairman of the transportation committee, hastened to point out a regulatory flaw to the national director. To qualify for priority in the age of rationing, a motor vehicle had to deliver to farmers raw materials, farm products and foods, and call upon farmers and pick up cream, eggs, and other agricultural products. There was one caveat: These vehicles could not deliver food to farmers for human consumption. Heinkel complained:

> *"…such a ruling would place those engaged in agricultural production on an equal basis with persons living in towns and cities and not require them to take time off from useful agricultural production to go to market."*
>
> —Fred Heinkel

Such a ruling, places farmers in a disadvantageous position and amounts to discrimination against them as compared with consumers living in towns or cities who are either in walking distance or can take advantage of public service facilities to buy food and other household necessities, while a farmer to do this must travel long distances and, if motor vehicle is used he has no way of replenishing tires and in most cases other facilities for transportation, such as horse and buggy, are not available because through years of use of automobile farmers have lost possession of horse-drawn transportation facilities.

Since trucks are permitted to deliver feed for live stock and pick up from the farmers agricultural products, it would appear that a rule could be made and a regulation adopted that would permit delivery of food to the farmers for human consumption and household necessities by trucks calling at their homes under present provisions and not result in any additional wear on tires, and such a ruling would place those engaged in agricultural production on an equal basis with persons living in towns and cities and not require them to take time off from useful agricultural production to go to market.

MFA's regular farm delivery trucks could pick up this slack, could, that is, if federal regulation were rational. The reply from Price Executive of Rubber and Rubber Products Section, Ben Lewis: "I appreciate your deep concern over the Tire Rationing problem as it affects the agricultural industry. Should there be an improvement in the rubber outlook, we will immediately relax the present restrictions." Period. Despite this up-close view of regulation, Heinkel continued to believe in the power of the federal bureaucracy. From his perspective, farmers would benefit more from a well-run federal policy than from a well-run business. All it took were the right people to design, and enforce, that policy.

MFA Oil Company supplied a full line of petroleum products as well as tires. Returning WWII veterans fueled a mechanical boom in the farm economy. MFA Oil developed multiple products to meet that demand.

Heinkel kept up a steady stream of correspondence with federal officials in his attempts to keep tires on and gasoline in farm vehicles, MFA milk trucks, farm supply trucks, feed trucks and exchange trucks. The MFA conventions were cancelled indefinitely in view of war shortages and rationing. They would be reinstated in 1944.

Simultaneously, Heinkel moved quickly in securing an oil refining plant at Chanute, Kan., to guarantee a steady supply for MFA Oil Company. In just four weeks, MFA received enough money from farmers at 5 percent interest to raise the $160,000 necessary to buy the plant. St. Louis Bank of Cooperatives supplied the balance. "A month later," wrote Derr in Missouri Farmers in Action, "the amount was oversubscribed, in spite of the fact that the farmers were at that time of year busy with their farm work, and MFA Oil Company officials, tank wagon men, and agents were busy supplying the farmers. Much of the money came in without solicitation." In addition to the plant, the purchase brought under MFA's control 115 miles of pipeline and a capacity of 1,500 barrels a day. By 1948, MFA Oil would buy yet another refinery: Delta Refining Company in Memphis, Tenn.

MFA Stores

Fortuna

Freeburg

Glasgow

Halfway

Hamilton

By 1944, MFA added a feed mill at St. Joseph, vastly increasing the reach of MFA feed products in northern Missouri.

In terms of business, net sales of all businesses operating under the MFA emblem delivered $104 million with savings of $2.5 million.

Producers Creamery continued its expansion. In this 1942 photograph, Producers Creamery of Kirksville added a second story to meet increasing demand.

Fred Heinkel was just getting started

By 1942, MFA membership exceeded 52,000. On the political side, Heinkel was named president of the State Highway Users Conference, a board comprising the state's largest highway users. In terms of business, net sales of all businesses operating under the MFA emblem delivered $104 million with savings of $2.5 million. According to Derr, "There were more than 325 units of the Association scattered across the state which figured in these totals." By 1944, MFA had added a feed mill at St. Joseph, solving what had been a problem in feed distribution in northern Missouri, in addition to two other feed mills, an egg drying plant, a poultry plant, a new condensed milk plant, a new cheese and whole milk plant, and three tire and recapping plants.

Heinkel still wasn't finished. His far-ranging vision spotted far more on the horizon.

Producers Creamery Company, organized in 1927, sold sweet cream, dry milk solids, condensed milk, evaporated milk and butter. In 1943, the creamery company had total sales of $6.5 million. (Clock donated by Stanley Cassidy)

PROUD PAST, BRIGHT FUTURE: MFA INCORPORATED'S FIRST 100 YEARS | 87

MFA-brand coffee, custom blended according to a secret formula, was highly popular throughout the trade territory. Whole beans were sent straight to MFA exchanges and groceries where grinders filled the stores with the smell of fresh coffee.

A year prior in 1943 at Heinkel's encouragement, MFA hosted 12 district branch presidents of the Bank for Cooperatives and the Central Bank for Cooperatives to tour the cooperative capital of the United States: Springfield, Mo. At that Springfield meeting, MFA officials guided the bankers through MFA's showcase of business structures in Springfield: Producers Produce Company (with a net worth of $458,000 and net sales exceeding $5 million); MFA Milling Company (with a net worth just shy of $1 million and net sales of $5.5 million); Producers Creamery Company (with a net worth of $1.27 million and total sales of $6.5 million); Producers Grocery Company (with a net worth of $37,000 and total sales of $1.3 million); MFA Hatchery (with a net worth of $20,000, total sales of $72,000 and 474,655 chicks hatched); Farmers Live Stock Commission Company (with a net worth of $76,000, 2,865 railcars handled, representing 15,865 cattle, 35,569 calves, 88,159 hogs and 53,455 sheep); Green County Farmers Sales Association (with a net worth of $80,700 and selling poultry, eggs, feeds, fertilizer, grain, berry crates, boxes, wire, nails, twine, salt and flour).

88 | FRED HEINKEL TAKES CHARGE

▲ Started in 1938, Producers Grocery Company operated out of a Producers Produce Company warehouse on Wall Street in Springfield, Mo. By 1978 sales exceeded $9 million and the grocery company was the only farmer-owned grocery company remaining in the United States. Seven years later, the company closed its doors, largely because, as Ray Young would note, most exchanges had exited the grocery business. The tremendous growth of supermarkets in the 1960s made the business model obsolete.

▼ MFA Grocery Company supplied exchanges with a complete line of grocery products. In 1942, MFA Grocery Company moved 440 train-car-loads of groceries to MFA exchanges and stores. This 1954 photograph is of the Aurora MFA store.

MFA continued its national impact. As Derr pointed out: "A compilation of sixteen regional cooperatives showed that [MFA's] 1943 business…was in excess of the total for all of these cooperatives, while the regional cooperatives had more than 2,500 separate cooperative associations and nearly 900,000 members."

At World War II's end in 1945, Fred Heinkel urged the membership to allow MFA to join the National Council of Farmer Cooperatives. NCFC would provide a national voice to cooperatives. MFA was the third largest cooperative represented. By 1948, Heinkel would be elected NCFC vice president and help drive national policy.

Also in 1945, MFA established a farm supply division through the efforts of a young MFA employee named Jack Silvey, and soon thereafter a seed division, feed division, plant foods division and grain division, a process eerily reminiscent of Howard Cowden's original plan. The first farm supply warehouse was built in Springfield. Others followed in Sedalia, St. Joseph and Maryland Heights. In that same year, MFA bought a packing company in Springfield.

PRODUCERS
PRODUCE COMPANY
SEDALIA, MISSOURI

*

Shell Eggs
Frozen Eggs
Dried Eggs
Live Poultry
Dressed Poultry

*

Wire Products
Galvanized Roofing
Asphalt Roofing
M. F. A. Paint

*

From Producer to Consumer

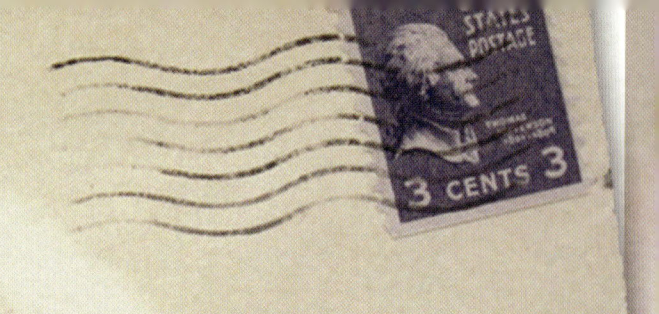

F. V. Heinkel
Missouri Farmers Ass'n.
Columbia, Missouri

Sedalia, M.O. — May 18, 1948 — 6:30 PM

M. F. A. Brands Are Dependable

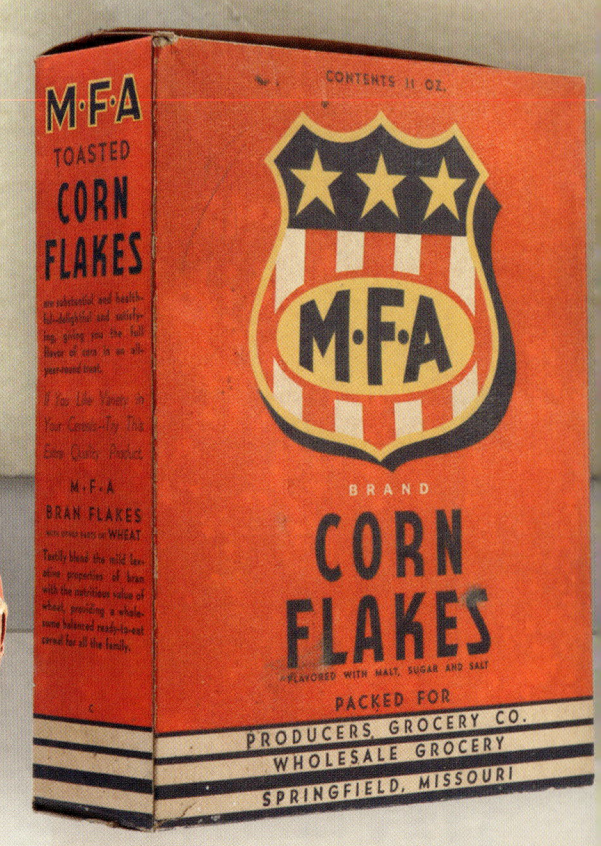

M·F·A

Producers Creamery Company

Daricraft
DAIRY PRODUCTS OF SUPREME QUALITY

NEW YORK OFFICE
100 HUDSON STREET
PHONES 5-2927
WALKER 5-2928
TELETYPE NY 1-942

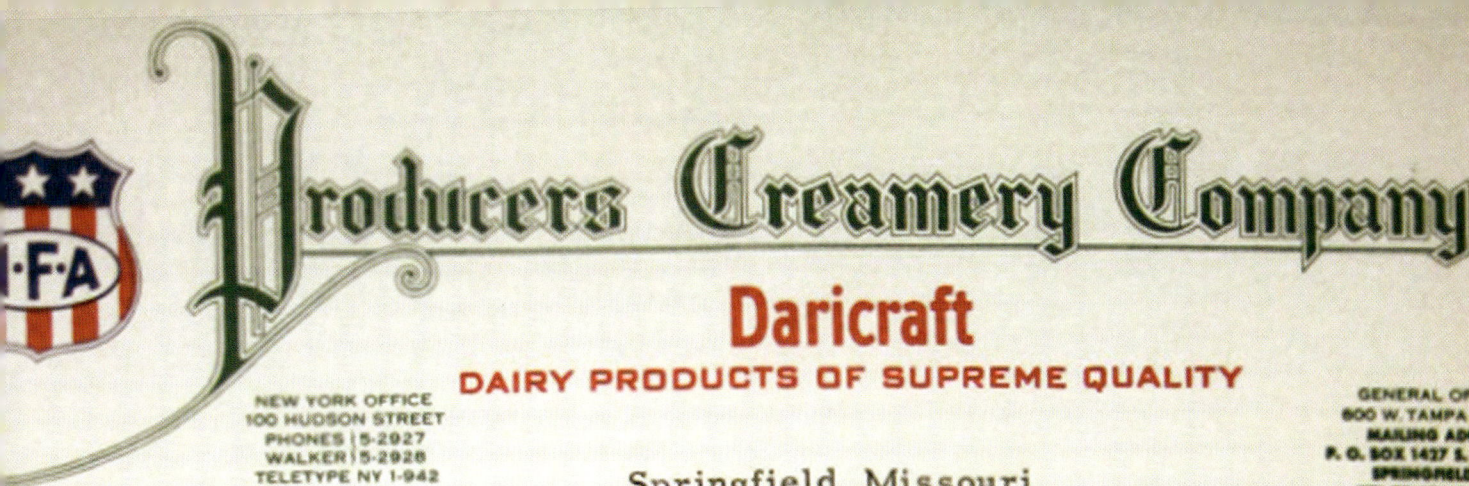

GENERAL OFFICES
800 W. TAMPA STREET
MAILING ADDRESS:
P. O. BOX 1427 S. S. STATION
SPRINGFIELD, MO.
TELEPHONE 2-7071
TELETYPE 502

Springfield, Missouri
April 26, 1954

Mr. F. V. Heinkel, President
Missouri Farmers Association
201 South 7th Street
Columbia, Missouri

Dear Fred:

I have a sister who directs food preparation of several hundred of the students at Missouri University. She informed me last week when I was there that they are using oleo at Crowder Hall.

I am wondering if the fact that Missouri is one of the outstanding dairy reason enough to serve butter

The flying bull

Production managers at MFA's Producers Creamery Company and MFA Milling Company couldn't help but be baffled by dairy trends in southwest Missouri in the mid-40s. Milk volume had been rising since 1940, but farmer profits weren't marching in tandem. Profits in the dairy business were dropping. Cow numbers increased, prices held steady but farmer profits declined. Genetics, they correctly concluded. Missouri's per-cow production had not increased for 15 years. Herd size averaged six cows.

The solution? Start a dairy stud farm on the outskirts of Springfield, home of Producers Creamery, to give MFA patrons access to world-class genetics. But how to deliver fresh bull semen over Ozark roads that, to be charitable, made timely travel difficult at best? The first attempt would be tightly scheduled appointments, fast cars and a network of inseminators.

In the heady days following victory in World War II, nothing was impossible and very few things seemed improbable. Artificial insemination had been introduced by the University of Missouri just a few years prior. The men who would create the MFA Artificial Breeding Association embraced its possibilities. Professional inseminators were identified and contracted in 30 strategic communities. Shelf life for fresh semen was approximately one day. That raised the stakes for delivery speed.

MFA purchased 30 of the finest dairy bulls obtainable (18 Jerseys, 8 Guernseys and 4 Holsteins) from all over the country, reaching even into Canada. According to The Missouri Farmer, "every bull has 600 pounds of butterfat per year

MFA purchased 30 of the finest dairy bulls available and began delivering bull semen to farmers in 32 counties. By 1949, MFA Artificial Breeding purchased airplanes to speed delivery and dropped vials of semen suspended under small parachutes. Success was almost immediate.

inheritance or better. One bull cost $4,000; three are grandsons of the world record Jersey cow; a three-quarter sister of one bull sold for $5,600 recently; four bulls are half-brothers and one is the son of the highest producing Jersey 2-year-old cow in Canada in 1942; and there are four grandsons of this cow's sire on hand which are yet too young for service."

Furthermore, while average southwest Missouri butterfat production was 200 pounds per year, The Missouri Farmer pointed out, "heifers of these bulls will produce 50 pounds more per year, and their daughters an additional 50 pounds per year. Thus the granddaughters of the dairy cows in that area will be producing 300 pounds of butterfat per year instead of the average of 200 pounds. You can multiply the price of butterfat by 100 and see what this increase will mean in farm income."

Better still, by relying on artificial insemination, farmers could get rid of existing herd bulls and replace them with an extra milk cow.

By 1949, MFA Artificial Breeding, frustrated with the slow pace of land delivery, purchased two airplanes, designed an impact-proof container and attached a small parachute to each container. By 1950, the planes were making 380-mile routes to more than 32 counties, swooping low over farms and dropping the packages that dangled under miniature parachutes. Success was almost immediate. Conception rates, hovering between 50 and 60 percent, confirmed the importance of aerial delivery.

On the fuselages of the airplanes (mimicking WW II aces), black bulls were painted under the caption, "The Flying Bull." MFA's impact on the genetics of Missouri's dairy industry continues to this day.

Investment continues

By 1946, MFA would make one of its wisest investments, putting the cooperative on the forefront of a crop gaining prominence throughout agriculture: the first cooperative soybean processing plant in the nation. The soybean processing plant was located at Mexico, Mo. MFA bought the facilities of an old mill and wooden elevator. The War Production board signed off on the deal.

With the completion of the plant, northeastern Missouri would experience a tremendous increase in both soybean acres and returns. Before the facilities had been built, soybean acres had been expanding in the northeastern counties. According to internal MFA documents, in a radio broadcast at the dedication, Heinkel pointed out that 14 counties within a 50-mile radius had produced 500,000 bushels valued at approximately $752,000 in 1941. After the plant was operational in 1946, farmers in those same counties

▲ MFA developed the first cooperative soybean processing plant in the nation and located it in Mexico, Mo., home county of William Hirth. The plant drove increased soybean prices for northeastern Missouri farmers.

▶ Fred Heinkel served as MFA president from 1940 until 1979. He was a business visionary. Under his watchful eye, MFA would build tremendous business volume through exponential growth and expansive political influence that grew national in scope.

produced 3.1 million bushels valued at $7.9 million. The plant stood front and center in the county seat of Audrain County, home county of William Hirth.

Missouri Governor Phil Donnelly appointed Heinkel to a commission to study the Tennessee Valley Authority and make recommendations for a Missouri River project. The U.S. Army Corps of Engineers was gathering data for what would become the Pick-Sloan program that's still in place today (with today's foolish and contradictory additions for recreation and environmental concerns).

The issue at the time was how to stop recurrent flooding in the river basin. Kansas City had faced deadly flooding repeatedly. Heinkel, ever the visionary, was impressed by the TVA. His practical nature saw potential in a similar system in Missouri. Heinkel dreamed big. He envisioned a Missouri Valley Authority. In Heinkel's mind, the river-containment system would include hydroelectric dams to power farms and fertilizer production facilities to feed farm production. Unfortunately, his was the minority view. His arts of persuasion fell on deaf committee ears. But Heinkel's proposal drew the attention of the Association for the Advancement of Science, officials of which asked him to speak to their organization in 1946 on "The Problems and Possibilities of the MVA."

At the 1945 convention, Heinkel announced the hiring of A.D. Sappington on a full-time basis as MFA's new counsel; creation of a new research department; organization of a new insurance company (MFA had had an insurance department since 1939) to open Jan. 1, 1946; and the new soybean plant in Mexico. The speed of these accomplishments was breathtaking—and exhilarating.

By the mid-40s, MFA and President Heinkel were starting to take another page from Howard Cowden's playbook. Interregional cooperatives (owned by two or more regional cooperatives like MFA) were proving a boon to the industry by providing quantity access to products through group purchasing. Like the regional cooperatives that formed them, interregionals focused on member benefits. MFA invested in Central Farmers Fertilizer Company, putting $100,000 toward capital stock for purchase of phosphate deposits. It was another long-term, strategic alliance that would pay huge dividends over the years. And it was the first in many additional interregional investments.

MFA Stores

Hartville

Herman

Higginsville

Huntsville

Hurley

By 1951, MFA plant foods division began offering free soil-testing bags, soil samplers and complete instructions at all exchanges.

MFA joined forces with the University of Missouri to offer complementary soil testing through Extension. Extension's mobile display was on hand at many MFA functions.

Market-research driven

The creation of MFA's new research department would pay dividends throughout the 40s and 50s. Heinkel, familiar with research staffs at the beck and call of congressmen and senators, saw immediate possibilities for MFA. "Before we determine where and when we are to establish a new cooperative," he wrote in The Missouri Farmer, "Dr. Haag [head of the new research department] will make a survey of the situation with a view to taking as much risk out of the venture as possible. In years gone by we have depended a bit too much on collective judgment and not enough on facts when establishing new cooperatives. … [M]istakes have been made in the past and might be made in the future if we are not exceedingly careful. In any event we shall all be better off if we eliminate all the doubts we can."

Simultaneously, Dr. Herman Haag was researching the potential for large-scale distribution of fertilizer. Haag, a professor of agricultural economics at the University of Missouri, had graduated from the university with the highest grades ever in the School of Agriculture. He earned his doctorate at Cornell. Haag brought a boundless work ethic and astounding practicality to his job. His reports pointed out fiscally sound opportunities over the years he was at MFA. Heinkel would push for action on many. The Mexico soybean-processing-facility purchase came about because Haag's market research convinced Heinkel of its practicality. Haag's research immediately led him to lend his support to Heinkel's idea of founding an MFA insurance company. And now plant foods were in the crosshairs.

Fred Heinkel and Herman Haag spoke as one on plant foods. Almost from the moment he ascended to the presidency, Heinkel had watched astounding increases in fertilizer use. In Missouri in 1941, farmers applied about 60,000 tons of fertilizer. By 1952, tonnage reached 800,000 tons—and would keep increasing. Because of Heinkel and Haag, MFA drove much of that demand.

At war's end, munitions plants easily converted from war production to domestic fertilizer production. Low-analysis, bagged fertilizer hit the market with big impact. At Heinkel and Haag's insistence, MFA focused on high-analysis. The advantage? Low unit cost and lesser weights to handle. Using Haag's research, Heinkel pushed for MFA-owned facilities.

Haag predicted Missouri-farmer use of more than 200,000 tons by 1946. According to Young's book, Haag conservatively estimated MFA could produce 25,000 tons of fertilizer a year by expending $140,000 and tying up $110,000 in working capital. By 1948, Heinkel's vision became reality. An MFA plant in Springfield churned out 8-24-8, 4-24-12 and 0-20-20, the highest analysis fertilizer ever sold in the state. Exceeding expectations, the plant manufactured and sold 40,000 tons in its first full year of operation.

By April of 1949, MFA's newly created plant foods division shipped from yet another facility, this one at Maryland Heights, a former munitions plant on the outskirts of St. Louis. Maryland Heights added 50,000 tons to MFA's production. The University of Missouri was on board with a complementary soil-test program through Extension. MFA managers asked farmers: "Why should you buy two tons of 4-12-4 when one ton of 8-24-8 will do the job?"

▲ By 1949, MFA had four full-time salesmen devoted exclusively to plant foods. Through demonstrations, a corn-yield contest and ready supply, MFA served farmers' growing demand for plant foods.

▶ MFA's corn-yield contest had two purposes: display the yield effects of the new plant foods products and supply a growing demand for corn hybrids. Corn yields would increase dramatically with the proper use of fertilizer. MFA, and farmers, benefited from both.

By summer of 1949, MFA had four full-time representatives pushing nothing but plant foods. They called on exchanges and lobbied managers on potential. Not all farmers were on board immediately. Many saw commercial plant foods as too much additional cost. But many more signed on enthusiastically—and dramatically increased crop production. More than 30 MFA exchanges moved product. Within five years, MFA would lead the state in construction of bulk plants to ease distribution bottlenecks.

By 1950, MFA installed the first anhydrous ammonia tank, probably the first in the state. Demand soared. By 1955, MFA had positioned 26 tanks and ramped up direct application. Simultaneously, MFA began its famous corn yield contest, with a primary purpose of promoting soil testing and stimulating proper use of fertilizer, new corn varieties and better crop management practices. In 1951 MFA plant foods began to offer soil-testing bags, soil samplers and instructions at all exchanges.

▼ By 1950, MFA installed its first anhydrous ammonia tank, probably the first in the state. Almost immediately, demand caused MFA to install 26 additional tanks around the trade territory. Anhydrous use was off and running.

Proud Past, Bright Future: MFA Incorporated's First 100 Years | 97

▲ MFA led the state in construction of bulk plants. In addition, MFA planned, built and ran fertilizer production facilities at Springfield, Joplin and Maryland Heights, on the outskirts of St. Louis. More than 300 MFA exchanges moved product to farmers. MFA produced a wide variety of plant foods, offered granulated pellets and encouraged use of 12-12-12.

▲ **Top Right:** MFA pushed soil testing to encourage the best use of plant foods. Soil testing became a large part of MFA's plant foods program from the early 1950s on.

Maryland Heights, locked onto Heinkel's vision, began production of 12-12-12 and began granulating pellets that stored well and did not lodge in spreaders. Between 1947 and 1951, MFA was just shy of doubling production and sales.

By 1953, MFA completed yet another facility, this one in Joplin. Joplin produced even higher analysis: 12-36-19, 14-14-14, 14-28-14 and others. By 1954, the first MFA bulk fertilizer plant opened at Albany. Two weeks later, Slater opened another. Before the next year ended, 10 MFA bulk plants were situated at strategic locations: Slater, Albany, Centralia, Gallatin, El Dorado Springs, Lockwood, Union, Butler, Rolla and California.

Jim Dissler, who for decades would deal with all aspects of fertilizing crops, hired on with MFA's plant foods division in 1955 at the height of the boom. He supervised much of the bulk-plant construction and oversaw many of the equipment upgrades. MFA even patented its own blender called the Newcomer (after the Newcomer Farm Club). Several still exist. Plant foods, which had initially begun shipping in 112-pound bags, moved to 80 and finally 50. Much of the plant foods was mixed with Aldrin for bug control.

Bulk plants were simple affairs with one-ton mixers. Trucks would back in; fertilizer would fall from overhead bins. MFA pushed hard for soil tests; MFA could mix almost any analysis farmers requested. Bulk plant office fixtures were No. 2 pencils, 10-key adding machines with hand cranks, and sometimes a desk and chair. Period. "There weren't five bulk plants in the state, when we started building them," Dissler said. Many of the plants featured plant foods in bins, many in bulk.

Much of the fertilizer production at large was anhydrous, sulfuric acid and rock phosphate to create ammoniated phosphate fertilizer. Springfield used a nitrogen solution and phosphoric acid to make a slurry. Potash would then be added to coat the granules. Each granule was complete: N, P and K.

Despite wide farmer acceptance of these products, one lagged behind: potash on alfalfa. Dissler came up with a scheme. He mapped out the biggest and best alfalfa producers in an area, loaded up a truck, and, unasked and uninvited, made a pass in each field in the shape of an MFA manager's initials. Of course, farmers noticed the initial—a foot taller than surrounding alfalfa and a far deeper green. Sales would skyrocket.

At the 1946 convention, Heinkel announced 100,000-plus members. The earned-membership plan was paying off in spades.

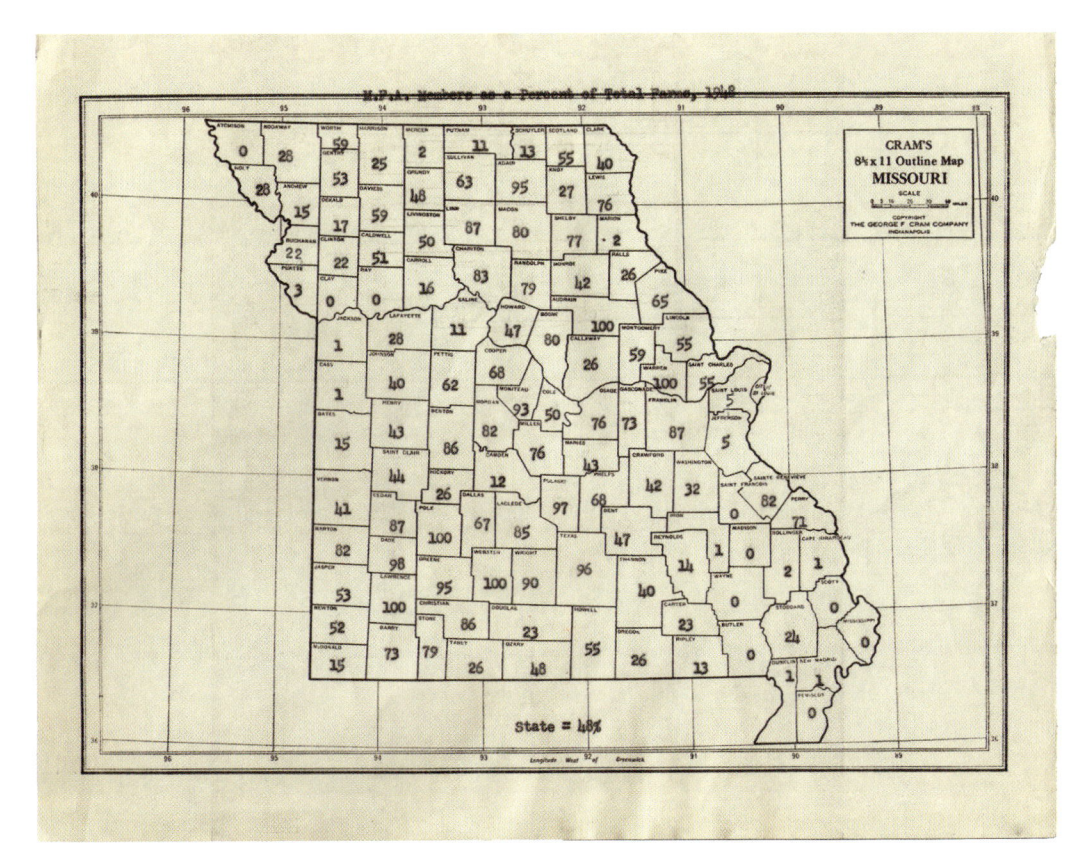

Heinkel would usher in a new earned-membership plan. Under this system, dues were paid by the MFA affiliate patronized by the member. When an individual conducted $25 worth of business at an MFA agency, the individual automatically became a member.

"There's a lesson here"

Plant foods was just one piece of MFA's multifaceted approach. Again, Heinkel thought large. Accordingly, in July of 1945 at his urging, the MFA board authorized an advance of $100,000 to establish an insurance company. According to A.D. Sappington (who would later become president of the insurance company), the money came from strategically withheld patronage. "I remember sitting in on board meetings when they were declaring what they called then patronage rebates," Sappington recalled in 1971.

Board members were arguing that the patronage should not be kept but returned immediately. As Sappington went on to explain, "Mr. Heinkel and some of the other farsighted board members said, 'No, we're going to need that money. We need to build up reserves and just allocate it to the exchanges so we can pay it back in a later year.' And because it was kept, the MFA, when they organized the MFA Mutual Insurance Company, had $100,000 to pay in as contributed surplus. And that's all the MFA Insurance Company started with. A hundred thousand borrowed money from the Missouri Farmers Association."

Originally, the board started out to organize the insurance company as a capital stock corporation with MFA owning the stock. According to Young's book, "Had this occurred, MFA would have been able to receive dividends and thus would have enjoyed substantial financial benefits and been assured of maintaining ownership and control. Because of income tax considerations, however, it was made a mutual company, with ownership and control in the hands of the policy holders. Financial benefits to MFA included payment for use of the MFA name, the coverage of a portion of MFA officers' salaries, the supplying of food at MFA conventions, and the loaning of

MFA Incorporated organized MFA Mutual Insurance Company with $100,000 as contributed surplus. Although the idea of an MFA insurance company had been around since the 1920s, the impetus came when MFA lost a court case and the contracted insurance company defaulted leaving MFA to foot the bill. "If we're going to have to pay both premiums and damages, there's no reason we shouldn't start our own agency," Fred Heinkel told Howard Lang. Lang would become the final president of MFA Mutual Insurance Companies.

money for building facilities." Fred Heinkel's lack of a number-cruncher's foresight in structuring the business would return with a vengeance.

Howard Lang, president of MFA Mutual Insurance from Sappington's 1979 death to the early 1980s, said the real impetus for Heinkel's zeal in forming the insurance company was a courtroom scene in the early 1940s. MFA had toyed with the idea of forming its own insurance company since the 1920s when managers had beseeched Hirth to form one. Lang was a prosecuting attorney. MFA was being sued by an individual whose car had been damaged by an MFA truck. MFA lost the case and was ordered to pay damages. MFA's insurer declared bankruptcy and stuck MFA with the bill on the accident.

After court adjourned, Heinkel approached Lang and graciously congratulated him on his successful prosecution. Lang remembers Heinkel saying to him, "There's a lesson here. If we're going to have to pay both premiums and damages there's no reason we shouldn't start our own agency. You interested in helping?" Lang declined at the time since he held public office. But several years later he would join MFA Insurance and gradually rise to president.

Jack Silvey was selected as the first manager of MFA Mutual Insurance Company. Silvey, another in a handful of MFA's excellent business executives, was brilliant by all accounts. He had been the first general manager of MFA Central Cooperative and built it into a profitable business, thus saving at-risk farmers exchanges. Silvey is credited with wise acquisitions and as the individual who worked to form the farm supply department as well as the first insurance department. But he was not and would never be a close Heinkel confidante.

Proud Past, Bright Future: MFA Incorporated's First 100 Years | 101

An award-winning existential fight

Throughout this initial building process beginning in the late 1940s and throughout the 1950s, Heinkel pushed back hard against opponents of cooperatives. That fight spanned a spectrum from farmer cooperatives to rural electric and telephone cooperatives.

An organization called the National Tax Equality Association fired barrage after barrage of charges against cooperatives. NTEA began targeting Missouri in earnest and gained national traction from businesses competing with cooperatives. Through a targeted campaign, NTEA sought to gin up opposition to cooperatives from local businessmen by insinuating co-ops were actively competing against local stores and were rewarded by government for doing so. NTEA's multi-pronged effort included constant lobbying of congressional leaders. In hindsight, the effort stands as testament to the effectiveness of the cooperative's market penetration. Heinkel took these battles personally. He stepped up MFA's congressional lobbying efforts, testifying on both the state and national stages.

"Certain misguided people have been waging a vicious attack upon cooperatives," Heinkel testified to Missouri legislators in 1945, "chiefly through disseminating misinformation in a whole sale manner." Heinkel told the politicians that of the 250,000 Missouri farmers, almost all belonged to either MFA, fire insurance co-ops, milk-marketing co-ops, Farm Bureau, electrical or telephone co-ops.

He intoned,

> What is the objective of these cooperatives in Missouri? They are organized for the purpose of increasing the farm income! And what do farmers do with this increased income? THEY SPEND IT IN TOWN [emphasis in original]! …And for your information, we of the Missouri Farmers Association feel that the entire Farm Credit Administration is the finest thing Congress ever provided for agriculture. It is founded upon the principle that it is a proper function of the Government to 'help people to help themselves.' We believe in this principle, and that's why we find no fault with the Federal Deposit Insurance Corporation which Congress established to help banks, or with the Reconstruction Finance Corporation to help business.

Heinkel, in that same speech, ended with:

> The M.F.A. is strictly a self-help organization. We have never received one cent from the Government. We have never accepted, nor been offered for that matter, one gratuity of any kind. If we have been successful thus far, it has been the result of hard work—this, plus the tears, heartaches, sweat and toil and sacrifices of thousands of Missouri farmers and their families. We repeat: The M.F.A.'s cooperatives exist for only ONE PURPOSE [emphasis in original]—to raise the level of living of farm families to a higher plane! Upon this premise we rest our case, and we will welcome any

An organization called the National Tax Equality Association began an attack on cooperatives that lasted for the decade of the 1950s. The association's attack was on the taxation status of cooperatives. MFA, through its association with the National Council of Farmer Cooperatives, pushed back hard through a series of print and radio ads and speaking engagements.

> *"The M.F.A.'s cooperatives exist for only ONE PURPOSE — to raise the level of living of farm families to a higher plane!"*
>
> —Fred Heinkel

102 | Fred Heinkel Takes Charge

MFA widely publicized its taxation payments, using this one of R.J. Rosier, MFA's corporate secretary, delivering a tax payment to the post office.

MFA Stores

Iantha

Iantha

Jamestown

Jasper

Jonesburg

inquiry any Congressman or Senator may care to make, and we invite you to pay us a visit and see Missouri cooperatives at first-hand for yourselves.

Heinkel would make a version of that speech over and over throughout the time period from the statehouse to the House Ways and Means Committee and the Senate Finance Committee in the nation's capital. Taxation, specifically Capper-Volstead, the co-op-enabling legislation, was at the heart of the issue. As Heinkel testified in the U.S. Senate in 1951,

> On this question of expansion by farmer cooperative associations, we farmers in Missouri have recently constructed two new fertilizer plants, in order to take care of our need for fertilizer and plant foods, which in the last four years have increased far beyond the capacity of all production facilities, including our own. We have built a soy bean processing plant and a seed processing plant, where none existed before, in order to serve our own needs. We have constructed three new milk plants in order to effectively

"Yes, in Missouri we have tried to expand and grow, and we will continue to do so, because we believe that the benefits flowing to Missouri farmers and to the entire economy of our state make our efforts worthy of the highest praise."

—**Fred Heinkel**

market for ourselves the increased volume of milk we are producing in southwest Missouri. In so expanding our operations, it was necessary that we farmers ourselves invest out of our pockets more than 2 ½ million dollars for these operations, and we were unable to expand them merely with the use of accumulated reserves. Yes, in Missouri we have tried to expand and grow, and we will continue to do so, because we believe that the benefits flowing to Missouri farmers and to the entire economy of our state make our efforts worthy of the highest praise.

There has been much loose talk to the effect that unless some kind of discriminatory tax is placed upon cooperatives, then all forms of business will convert to the cooperative method of doing business in order to escape the payment of federal income taxes. Such a charge is fantastic. General Motors, United States Steel, or duPont Company would not give a second thought to accruing and paying all of their net profits over to the people who ultimately buy and use their products. They are in business, and rightfully so, for the purpose of making a profit for the owners of those businesses, namely, the stockholders. Farmer cooperatives, on the other hand, are in business for the purpose of enabling their patrons, the farmers, to increase their income out of their own farming business.

MFA's information department produced voluminous quantities of print ads, radio spots and brochures detailing the breadth of taxes paid by cooperatives.

For the next two decades, MFA, specifically through the efforts of Fred Heinkel, was at the forefront of cooperative battles. Heinkel, seeing whole cloth, testified not just on agricultural cooperatives but on farmer-owned cooperatives in any form. In

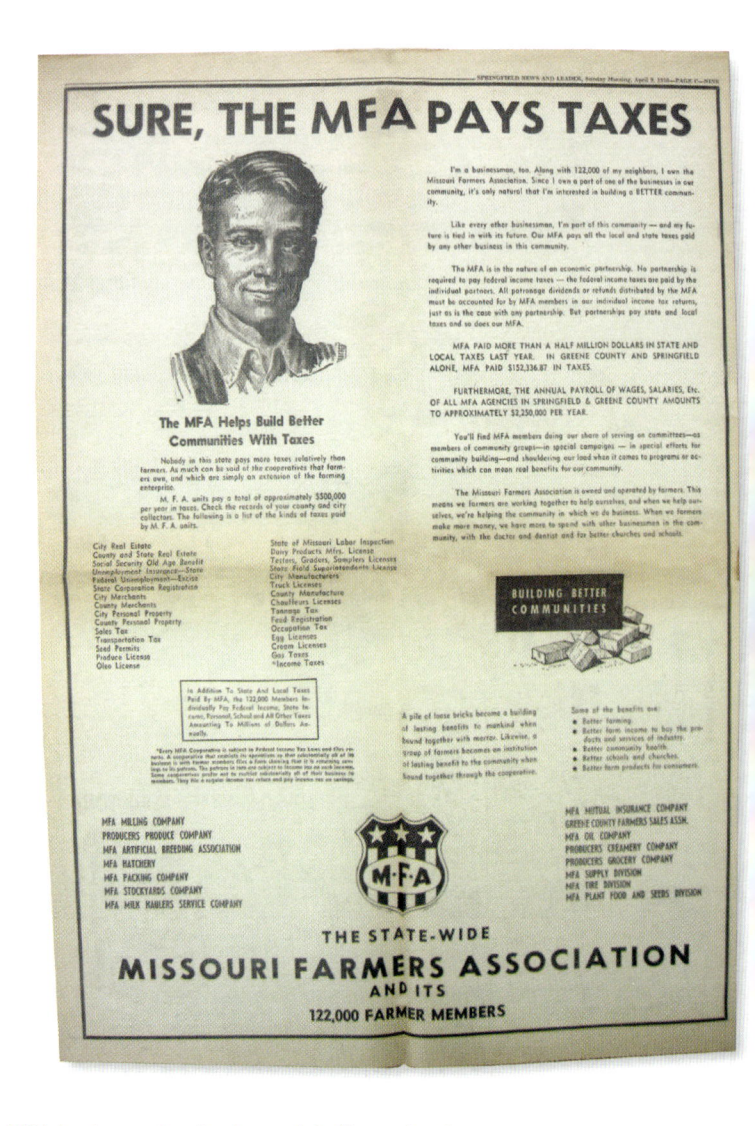

MFA took great pains to explain the voluminous taxes MFA paid in national, state and local venues.

104 | FRED HEINKEL TAKES CHARGE

forceful testimony, he focused on electric and, later, telephone co-ops as well, even pledging in MFA's convention resolutions vigorous support of "all state and national legislation which will insure the continued progress of R.E.A. [Rural Electrification Administration] cooperatives, and we will oppose any legislation or administrative policy…which will hinder the growth or operation of R.E.A. …"

Fred Heinkel knew farm life in the days before electricity. He led opposition against selling electric co-op assets to private industry and against diverting power-transmission lines from public dams to private utilities at the expense of co-ops. In fact, over the years, Heinkel would become so effective in his support of rural electric issues that individual electric cooperatives passed resolutions honoring his efforts, as did the Missouri state association of electric cooperatives as well as the National Rural Electric Cooperative Association Region VIII (Missouri, Arkansas, Oklahoma and Louisiana).

Fred Heinkel's spirited defense of the cooperative form of business and his non-stop congressional actions in moving that argument forward led several rural electric organizations to award him their highest honors.

By 1970, the NRECA would present Heinkel with its Distinguished Service Award. Robert Partridge, general manager of that national organization, wrote Heinkel:

> This award, which is presented to the American whom the Board feels has given the most outstanding service to the rural electric and public power programs, as well as to the total development of rural America, has been given only eight times. Those who have been thus honored in the past include George Norris [Hirth's old friend], Franklin Roosevelt, Sam Rayburn, Morris Cook, Harry Truman, Leland Olds, Jerry Voorhis, and Clyde Ellis. …Fred, I want you to know that I am both professionally and personally proud of the honor the Association is according you, and this letter is written with a great deal of joy.

"We have no apologies to make"

Historic drought burned crops in 1953. Farmers looked at cattle herds and empty haylofts and despaired. In the Ozarks, the area of the state hit the hardest, cattle grazed on brush or nothing. Fred Heinkel spent the sweltering days in Jefferson City, lobbying his friend, Gov. Phil Donnelly. Heinkel wanted the governor to call a special session of the General Assembly.

Heinkel's proposal? If Missouri would set a ceiling on hay prices (to discourage panic buying and speculation) and appropriate $6.5 million for rail transportation, MFA would handle finding, organizing and distributing hay to farmers at cost.

Heinkel's argument was straightforward. MFA had the national contacts and staff to find hay and get it loaded. Plus, as the largest single customer of Frisco Railroad and the second largest shipper on the Missouri Pacific, MFA had influence. Fred Heinkel used that influence to get both railroads to guarantee emergency rates of $206.78 rather than standard fare of $413.56. Railroads had already agreed to guarantee those rates, but only until the end of December 1953.

Donnelly complied. The General Assembly met. The package sailed through. As Heinkel would write to MFA managers on Nov. 16, 1953, "The recent special session of our General Assembly was an event unique in American history. It is the first

Fred Heinkel (center) presented his friend Gov. Phil Donnelly (left) and Clell Carpenter, Missouri secretary of agriculture, with awards of appreciation for their efforts in coordinating the hay lift of 1953. MFA would coordinate distribution of 480,000 tons of hay during the historic drought. Heinkel would form a long association with Carpenter, who, in turn, coached Heinkel in political activities on both the state and national levels.

time any state legislature ever met in special session and appropriated six and one-half million dollars for paying the freight on shipments of hay from outside the state for the purpose of helping farmers preserve their livestock herds."

As Heinkel would further point out, he offered MFA's free distribution services with the knowledge "that every M.F.A. Manager and Director, being aware of the deep tragedy suffered by Missouri farm families because of the drouth, would want to cooperate fully to the end that further dispersal of our beef and dairy herds might be averted; and I was therefore confident that you would all back me up."

John F. Johnson, the man Hirth had finally found to make a success of MFA Milling Company, took charge. Eventually, MFA Milling Company distributed 480,000 tons of hay, mostly from the Dakotas, but some from other northern states and even the East. MFA made life-long customers out of affected farmers. Twenty dollars a ton was the ceiling price. But the maximum paid for alfalfa would be $19 a ton and $17 on wild, prairie grass.

"I will sleep tonight," said Wilber Stigall of Hartville in an MFA news release of the time. He picked up one ton of "wild grass" hay in Mansfield. "I paid $15.50 a ton for this hay, and it looks to me like it is good quality. I was out of hay. I think it was a life-saver to all of the farmers—especially little farmers like me."

By the time the program ended on Dec. 23, 1953, railroads resumed their standard price schedule. Competitors complained that MFA had tied up the railroads the entire period and they were shut out. Johnson took to the Springfield newspaper to answer charges in an ad. He'd quit counting railcars at 16,000, so he didn't have an exact number to show those competitors. He was fresh out of apologies, too. "Probably this tremendous movement of hay in this short period of time was responsible," he admitted. But he also added, "we have no apologies to make."

The hay lift, as it was called, contributed one additional item to MFA as an organization. A man named L.C. "Clell" Carpenter was the state commissioner of agriculture at the time of the event and worked closely with Heinkel in administering the program. When Carpenter's term expired, Fred Heinkel hired Carpenter as MFA's

106 | FRED HEINKEL TAKES CHARGE

legislative representative in both Jefferson City and Washington, D.C. He would become Heinkel's right-hand man.

Carpenter, alternately praised and vilified by observers, was a political professional. He had run for governor and served as Missouri's highest agricultural official. He had been both state and national head of Farmers Home Administration. He knew politics inside and out. Over the course of the next two decades, Carpenter would tutor the politically precocious Heinkel, refine his political skills and help launch Heinkel's most effective endeavors on a national stage. Carpenter would also launch MFA (and Fred Heinkel) as a more effective and powerful force in both state and national politics.

Clell Carpenter would try his hand in MFA's business operations for one year. He began by organizing and implementing MFA Livestock Association. At the end of one year of large operating losses, he threw in the towel and went back to his field of expertise: politics.

"I'm sure you recall your inaugural address"

From Hirth to Heinkel, MFA had maintained close ties to the land-grant university in Columbia. The University of Missouri was a ready source of agricultural professionals in all areas pertaining to crops and livestock. Professors constantly provided how-to and why articles for The Missouri Farmer.

So when university officials approached Fred Heinkel in 1950 about political efforts to influence the MU Board of Curators to move medical training to Kansas City and expand it to four years at the expense of the Columbia program, Heinkel took the issue to heart. At this point in time, the University of Missouri had only a two-year medical program.

Heinkel pointed his thorough research director, Herman Haag, toward the issue of healthcare, specifically rural healthcare. Haag, with his usual precision, designed a massive study that turned national in scope.

Haag researched these astounding numbers of categories: hospital beds per 1,000 population; doctors per 1,000 population; the number of medical doctors (both MDs and ODs); hospital beds per 1,000 population districts (St. Louis, Kansas City, Springfield, Kirksville, Columbia, St. Joseph and Cape Girardeau); population per medical doctor by area; population per doctor in rural and urban areas; where current doctors were trained; relationship of size of largest urban center to number of residents per doctor; location of schools training medical doctors licensed from 1930 to 1949 and practicing in Missouri towns of less than 10,000 population as of Jan. 1, 1950; number of Missouri medical students enrolled in each school 1946–50; medical students per 100,000 population 1946–50 by states nationally; number of medical schools and enrollment of medical schools in each state; hospitals approved for internships by the American Medical Association and available; hospitals with internships in cities where university is located; and number of internships in hospitals approved for such training by the American Medical Association by state in 1950.

The results of Haag's research raised alarm statewide. The results weren't pretty. One district had 4,561 people per doctor, another 1.86 beds per 1,000 population. As you'd expect, Kansas City and St. Louis had twice as many doctors and hospital beds per 1,000 population as the remainder of the state.

Newspapers picked up the banner and ran. Heinkel was quoted everywhere. The St. Louis Post-Dispatch quoted Heinkel saying the study shows the state medical school should be located in a rural area. Going on to quote Haag, the newspaper printed, "It is readily apparent that the facilities for adequate health programs involving prevention of illness, as well as the adequate care of the sick are badly deficient in the rural areas of the state. Missouri has a rural health problem."

In the MFA archives is a series of letters between Heinkel and Missouri Governor Forrest Smith. Elected in 1948, Smith served one term. He and Heinkel were no strangers. Smith was a Democrat and Heinkel had watched his campaign closely. On March 15, 1952, Heinkel wrote to Governor Smith encouraging his support for appropriation of $6 million for a four-year medical school. "Since in a few days the $6,000,000 appropriation for a medical school will likely be

MFA's Role in the University of Missouri Medical School

Dr. Hugh Stephenson, legendary MU physician, credited MFA President Fred Heinkel as the driving force behind the decision to fund a four-year medical school and locate it in Columbia. This series of letters testifies to Heinkel's political finesse.

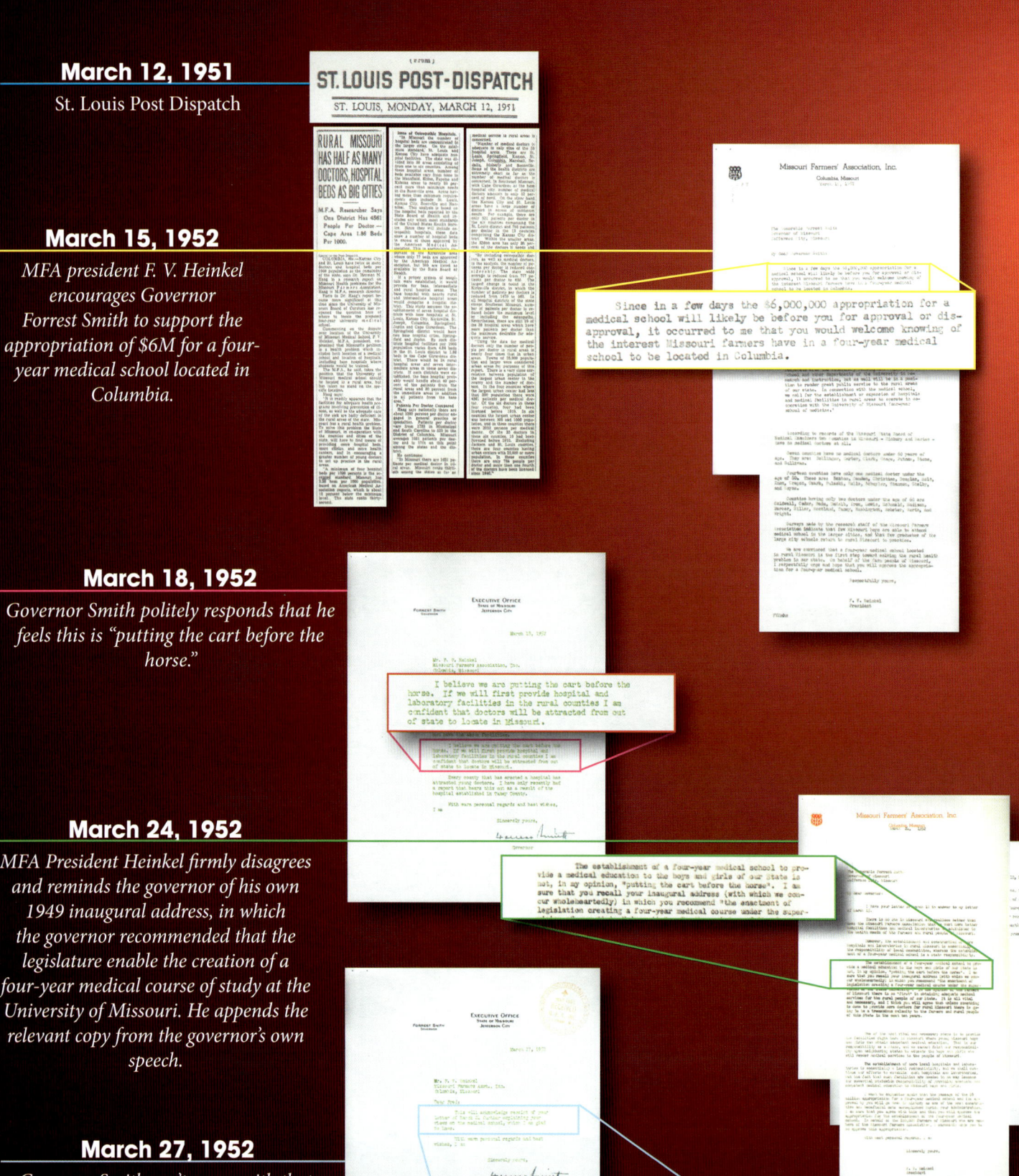

March 12, 1951
St. Louis Post Dispatch

March 15, 1952
MFA president F. V. Heinkel encourages Governor Forrest Smith to support the appropriation of $6M for a four-year medical school located in Columbia.

> Since in a few days the $6,000,000 appropriation for a medical school will likely be before you for approval or disapproval, it occurred to me that you would welcome knowing of the interest Missouri farmers have in a four-year medical school to be located in Columbia.

March 18, 1952
Governor Smith politely responds that he feels this is "putting the cart before the horse."

> I believe we are putting the cart before the horse. If we will first provide hospital and laboratory facilities in the rural counties I am confident that doctors will be attracted from out of state to locate in Missouri.

March 24, 1952
MFA President Heinkel firmly disagrees and reminds the governor of his own 1949 inaugural address, in which the governor recommended that the legislature enable the creation of a four-year medical course of study at the University of Missouri. He appends the relevant copy from the governor's own speech.

> The establishment of a four-year medical school to provide a medical education to the boys and girls of our State is not, in my opinion, "putting the cart before the horse". I am sure that you recall your inaugural address (with which we concur wholeheartedly) in which you recommend "the enactment of legislation creating a four-year medical course under the super-

March 27, 1952
Governor Smith can't argue with that. Within the year, a $13M appropriation for the school is approved. Construction begins in 1953.

> This will acknowledge receipt of your letter of March 24 further explaining your views on the medical school, which I am glad to have.

> *"There is no one in Missouri who realizes better than does the Missouri Farmers Association that we must have better hospital facilities and medical laboratories to administer to the health needs of the farmers and rural people of Missouri…"*
>
> —Fred Heinkel

before you for approval or disapproval, it occurred to me that you would welcome knowing of the interest Missouri farmers have in a four-year medical school to be located in Columbia."

Heinkel proceeded to quote an MFA resolution passed the previous August at the MFA convention that supported a four-year medical school at the University of Missouri. "According to records of the Missouri State Board of Medical Examiners," Heinkel further informed the governor, "two counties in Missouri—Hickory and Maries—have no medical doctors at all. Seven counties have no medical doctors under 60 years of age. …Fourteen counties have only one medical doctor under the age of 60."

Heinkel concluded with: "We are convinced that a four-year medical school located in rural Missouri is the first step toward solving the rural health problem in our state. On behalf of the farm people of Missouri, I respectfully urge and hope that you will approve the appropriation for a four-year medical school."

In reply on March 18, the governor balked, saying: "I believe we are putting the cart before the horse. If we will first provide hospital and laboratory facilities in the rural counties I am confident that doctors will be attracted from out of state to locate in Missouri."

On March 24, Heinkel firmly disagreed. "There is no one in Missouri who realizes better than does the Missouri Farmers Association that we must have better hospital facilities and medical laboratories to administer to the health needs of the farmers and rural people of Missouri. However, the establishment and construction of more hospitals and laboratories in rural Missouri is essentially the responsibility of local communities, whereas the establishment of a four-year medical school is a State responsibility. The establishment of a four-year medical school to provide a medical education to the boys and girls of our State is not, in my opinion, 'putting the cart before the horse.'"

To make the point even more clear, Heinkel appended a copy of Gov. Smith's inaugural address that reads, "I recommend the enactment of legislation creating a four-year medical course under the supervision of our State University…" Check.

Gov. Smith, finding no wiggle room, thanked Fred Heinkel for his views and signed off with "warm personal regards and best wishes." Checkmate. Within the year, Gov. Smith signed a $13 million appropriation for the school and hospital. Construction began in 1953.

Letters from university officials poured into MFA from a variety of people, including the president of the Board of Curators, Powell McHaney, and University President Frederick Middlebush. They had one central theme: thank you and thank the Missouri Farmers Association. In 1982, Dr. Hugh Stephenson, chief of staff at MU's Health Sciences Center, would write to thank Heinkel: "I hope that you will always feel proud of the major part that you played in bringing this about." Stephenson was one of the young doctors lobbying for the four-year medical school in the early 1950s.

MFA took over the entire area under the grandstands at the state fair to promote different MFA services and products. All MFA affiliates joined in the effort to support the state fair and display MFA's wide-ranging offerings.

To acknowledge that gratitude, those same men would use their influence to get Fred Heinkel appointed to the MU Board of Curators in 1953. Fred Heinkel had a sixth-grade education. But he was a man of accomplishment. He enjoyed national influence.

Revolutionizing hog marketing

In 1956, Don Larsen of the New York Yankees was hard at work pitching a perfect game in the World Series, beating the Brooklyn Dodgers 2 to 0. Dr. Herman Haag had almost as formidable a challenge. He was once again pitching a new MFA operation: this one a coordinated attempt to market livestock—hogs, specifically.

MFA since its beginning had focused on livestock marketing, first at the local level, then at the state and regional with Hirth's efforts in formation of Farmers Livestock Commission Company in St. Louis as well as Kansas City and St. Joseph. This time, however, hogs were front and center. Haag set as his objective an interior market catering exclusively to member needs. Haag polled members to determine their current marketing practices. Haag's report would serve as a guide "in drawing plans for a livestock marketing structure which will benefit all farmer-producers in Missouri, large and small producers alike, by making it possible for them to produce types of animals that meet high standards of quality and grade; and to market their livestock at the highest prices obtainable for the grade offered for sale."

By 1957, the project was up and running. It would be incorporated as MFA Livestock

Association, Inc. on April 18, 1958. As Heinkel wrote in the April 1958 issue of The Missouri Farmer, "This past month the Directors of your MFA Livestock Marketing Association (successor to Farmers Livestock Commission Company) launched a program which offers farmers the opportunity to take a real part in the marketing of their livestock. Briefly it consists of the establishment of a number of local yards—where farmers can be assured of fair weights, grades, and prices for their hogs—and the setting up of a sales office to sell these hogs."

Announcing was one thing. Implementing was another. No sooner had Heinkel announced the program than negotiations collapsed for a planned purchase of 14 existing country buying stations in northern and central Missouri. Because of momentum generated by the official announcement, speed was of the essence. As mentioned previously, Clell Carpenter had been designated general manager and would soon discover he had no stomach for market stress.

By May 1958 other facilities had been located and bought at Chillicothe, where the new association set up its first hog-buying station. By July, a second station was purchased, this one at Marshall. The two stations averaged 252 hogs a day.

Heinkel received a letter questioning the need for the new association from William J. Campbell, a Carrollton MFA member. In response, Heinkel wrote:

> My farm is located about thirty miles west of St. Louis and I have always sold all of my livestock on the terminal market at East St. Louis, Illinois through our firm, The Farmers Livestock Commission Company. But during the past fifteen or twenty years, contrary to your views and mine, thousands of Missouri farmers have decided not to sell their hogs on the terminal markets, and although in many instances they are merely surrendering their hogs to a buyer who has only one outlet, they continue to market in this fashion in preference to going to the terminal markets.
>
> Confronted with this situation, we have felt that it would be advisable to establish interior markets that would be owned, controlled, and operated by and for farmers. …In the case of our interior markets, the farmer is in a position to know approximately what he will receive for his hogs before he loads them in the truck. On the other hand, when he dumps them on the terminal market, he risks hitting a day when there are more hogs than the market will consume. And when that happens, the price collapses.
>
> Without a doubt the central markets will remain in operation, and I am inclined to think their prices will be stronger because the interior markets will help these terminal markets to avoid the heavy runs of hogs that are so detrimental to the farmer's receiving a proper price for his hogs.

MFA Stores

Koshkong

Laddonia

LaGrange

Lancaster

The MFA Livestock Association was off and running. New facilities opened that same year in Boonville, Cole Camp, Salisbury and Trenton. The initial facility purchase, which had broken down at the last minute, was again in play. This time, MFA inked the deal for a marketing system headquartered in Marshall. By late September, MFA found itself owner of additional facilities at Marshall, Sedalia, California, Stanberry, Tipton, Fayette, Blackwater, Slater and Bucklin.

By the end of 1958, after six months of operations, MFA Livestock Association had purchased 82,912 hogs, sold them to 26 packers and lost $51,342.20. Clell Carpenter had had enough. He resigned to return full time to public affairs in the political arena. The association lost $93,657.80 in 1959. The situation was critical enough that the St. Louis Bank of Cooperatives advised pulling the plug. Bankers would not renew a line of credit.

Heinkel stalled for time. A round of budget cutting ensued with layoffs and eliminations at non-performing locations. Volume began picking up. By May 1960, the association was able to announce slight savings for the month. The association's

This photograph from the late 1950s shows local affiliate MFA Cooperative Association of Salem.

▲ The MFA Tel-O-Auction serves as an example of MFA's willingness to innovate in all areas of agriculture. The service, by conference call, linked buyers from across a four-state area and brought better bids to swine producers throughout MFA's trade territory.

▶ MFA celebrated its 50th birthday in 1964 with festivities at the Newcomer Schoolhouse.

fortunes began to improve. By May another facility opened in Macon. Slight growth continued in 1962. And fortunes improved significantly when the Livestock Association began to focus on feeder pigs.

By 1964 the Association had a strong hog marketing organization in northern and central Missouri and a small, but growing, private treaty feeder-pig program in south-central Missouri. Simultaneously, marketing efforts expanded into Princeton. By now, the Livestock Association had ensconced itself in the middle of the existing feeder-pig markets in the state. Profitability followed. As did innovation. By 1965, MFA Livestock Association officials, after studying a unique system developed in Virginia, were ready to implement innovation.

On Dec. 2, 1965, 1,573 feeder pigs were grouped in key Missouri locations. A conference call connected buyers across a four-state area. The buyers had been informed of the process and guaranteed accurate description of the animals. An unqualified success, MFA Tel-O-Auction launched a revolution and immediate growth. New barns opened in Alton, Westphalia, Taneyville and Cabool.

PROUD PAST, BRIGHT FUTURE: MFA INCORPORATED'S FIRST 100 YEARS 113

Through his efforts in promoting farmer cooperatives, Fred Heinkel made a series of important national contacts almost from the day he assumed the presidency of MFA. He was the friend and confidante of national figures.

★★ Chapter 6 ★★
Political Distractions

"Sometime when you are in Washington"

Although Fred Heinkel had been a presence on the political stage since the day he'd assumed the helm of MFA, the late 1950s and early 1960s marked a noticeable shift toward the political arena, probably thanks in no small part to the masterly efforts of L.C. Carpenter. That presence almost exclusively involved Democrat politics. As Heinkel would himself note, MFA could make more money for farmers in one piece of legislation than in any other business endeavor. Washington policy figured mightily in agricultural profitability. And Heinkel could lift above his weight.

Heinkel accepted presidential commissions, starting years before with his Truman appointment to the Missouri Basin Survey Commission. He served as a member of a six-man group overseeing USDA's insect and plant disease control programs and as chairman of the national advisory committee on feed grains and wheat.

On the state front, he was front and center in political circles and served a term on the University of Missouri's Board of Curators in the mid-1950s.

Testifying to Heinkel's national political power, in 1956 U.S. Senator Stuart Symington of Missouri wrote to Heinkel, "As a new member of the committee, I am trying to learn what it is all about and do appreciate very much your suggestions and help." The following year, Symington further inquired: "I would appreciate your reading this talk [by Utah Senator Watkins] and giving me your frank opinion about it at earliest possible convenience. Your thoughts would help my position with respect to this bill when it comes before the Senate Agriculture and Forestry Committee."

Simultaneously, Heinkel started a whirlwind of speaking engagements. He spoke to state and national agricultural groups and at farm forums, to feed-grain organizations and advisory councils. He presented MFA positions to the Missouri Senate and General Assembly and to the U.S. Senate and House of Representatives on multiple occasions.

By 1962, when Fred Heinkel called for a conference of national farm leaders to consider national farm legislation, 127

Fred Heinkel, starting in the 1950s, was a regular presence in the nation's capital. He served on committees, commissions and boards dealing with national issues. (Western Historical Manuscript Collection)

people responded from 50 organizations in 22 states. Response was testament of Heinkel's political stature. During the same year, Heinkel would be elected vice president of the National Council of Farmer Cooperatives.

Although politics had always been part of MFA's conventions with resolutions being front and center, now political resolutions dominated. The 1962 report of the resolutions committee stated MFA positions on: national farm programs, Agricultural Stabilization and Conservation Service farmer-committees, international trade, federal marketing of farm products, tax reduction, taxation of mutual insurance companies, percentage depletion for farmers, the food for peace program, food for defense, food stamp plan, Peace Corps, farmer credit requirements, Farmers Home Administration, Banks for Cooperatives, public education financing, rural electric power, interstate trade barriers, rural development, natural resources, Missouri River navigation, Missouri economic study, Missouri constitutional convention, gasoline tax, county zoning, radio service to farm areas, roads and highways, animal diseases and cooperation among agricultural agencies.

For those subjects not covered, this catchall was added: "Policy Determinations. We recognize the inability of this Convention to adopt resolutions covering all phases of our farm policy or matters affecting farmers, particularly during periods such as the present when we are faced with increased world tensions and rapidly changing domestic affairs. In the absence of a specific resolution covering any specific item of farm policy or matters affecting farmers,

MFA Oil gas stations dotted the countryside throughout the sales territory. The globes on top of the pumps sell at astronomical prices in today's antique marketplace.

we do authorize the board of directors of Missouri Farmers Association to promulgate and adopt a policy relating to the same, which, however, shall not be inconsistent with the programs adopted by these resolutions."

At the 1960 MFA annual convention, the entire MFA organization presented Heinkel with an award for "extraordinary and outstanding service." In the text of the award is recognition of Heinkel's national standing: "because of your demonstrated leadership among farm groups throughout the length and breadth of this great Nation in the cause of a prosperous agriculture[.]"

That same year, Fred Heinkel received a letter from another U.S. Senator, this one named John F. Kennedy. "I have heard many fine things about you," wrote Kennedy, "about your knowledge of agriculture, and about the eminent position you hold in the minds of all who are interested in and knowledgeable about agricultural policy in our country…I want you to know, too, that we will be counting on you to consult with and advise us on the formulation of agricultural policy. I hope you will attend our meeting of the Midwestern states in Des Moines, Iowa, on August 21st, and sometime when you are in Washington I hope you will give me a call. It would be a pleasure to see you."

A national figure

Fred Heinkel hesitated briefly, cursed silently and boarded an airplane for the first time in his life. Even at this point in the Space Age (1961), airplanes didn't exactly impress him. Heinkel, after all, was widely known for having both feet planted firmly on the ground. As president of the politically potent 155,000-member Missouri Farmers Association for 21 years, he'd been well served by trains, even on cross-country trips. But this was no leisurely journey: U.S. President John F. Kennedy had requested an audience with Heinkel the next day, and a train just wouldn't do.

▲ Fred Heinkel (second from President Kennedy's right) was a member of the president's task force on farm surpluses in the early 1960s. After Kennedy's assassination, Heinkel would work closely with President Lyndon Johnson. (Wide World Photo)

◀ In 1966, then U.S. Senator Hubert Humphrey attended and addressed the MFA annual convention. He is shown here shaking hands with Ray Young, executive vice president of MFA Oil Company. In 1961, Humphrey recommended Heinkel to President Kennedy for the U.S. Secretary of Agriculture appointment over Orville Freeman, who was from Humphrey's own home state of Minnesota.

Kennedy was finishing the political appraisal process inherent in choosing a U.S. Secretary of Agriculture. The two finalists were Heinkel and a young, recently defeated governor from Minnesota named Orville Freeman. Oddly, Senator Hubert Humphrey of Minnesota would recommend Heinkel to the Kennedys over his home-state politician. Humphrey was a Heinkel fan.

In Washington, a member of the president's staff chauffeured Heinkel to Kennedy's Washington home. A large group of reporters gridlocked the front entrance, so Heinkel's car moved on down the street and slipped up a connecting alley. Heinkel was escorted through the backdoor of Kennedy's home, through the kitchen and into the study where President Kennedy was on the phone. Shortly thereafter Robert Kennedy walked in and the three men moved to a parlor with a fireplace. Heinkel and the Kennedys talked agricultural policy.

Soon thereafter, Freeman got the nod in spite of (some say because of) Heinkel's more extensive experience. Rumor had it Heinkel was JFK's pick, but Attorney General Robert Kennedy was wary of Heinkel's age (64) and lack of formal education (sixth grade). President Kennedy would tell Missouri Senator Stuart Symington that Heinkel was just too old for the youthful image the administration wanted to present.

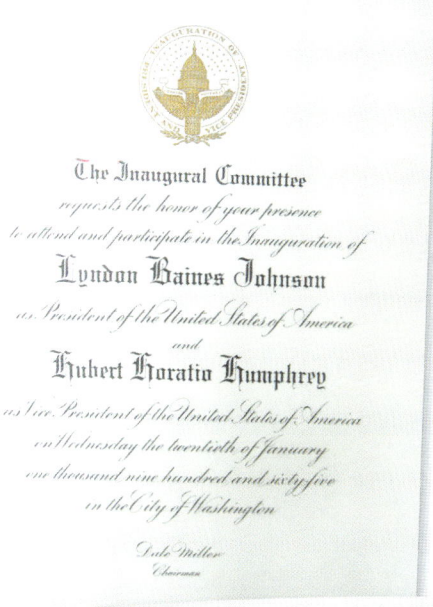

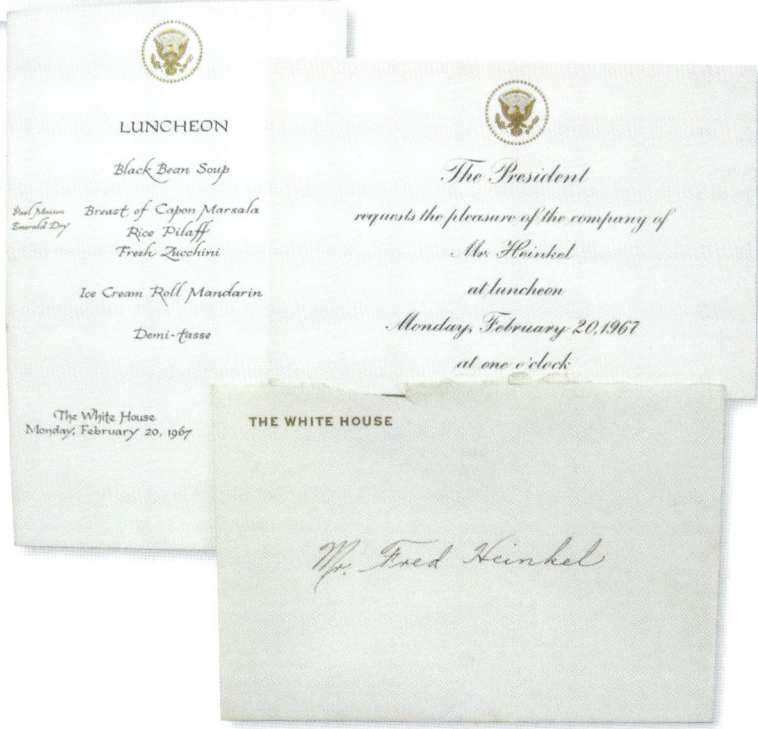

MFA Stores

Leonard

Lewistown

Lincoln

Lohman

Marshfield

Heinkel was a member of President Johnson's National Agricultural Advisory Commission. Heinkel is pictured on the left side of the photograph.

Fred Victor Heinkel in 1961 was 64 years old. He was at the pinnacle of his national political power as well as his administrative power as president of MFA. He would continue to be effective on the political scene for the next 18 years. He'd attend the inauguration of Kennedy and Johnson, as well as, four years later, the inauguration of Johnson and Humphrey.

After Kennedy's assassination, President Lyndon Johnson would continue the relationship with Heinkel. By 1964, Johnson was telegramming Heinkel he'd commissioned him to the National Agricultural Advisory Committee. President Johnson would write Heinkel in 1965, outlining agriculture's place in his vision of a "Great Society."

When MFA celebrated its 50th anniversary on March 14, 1964, Fred Heinkel was 67.

120 | POLITICAL DISTRACTIONS

Taking a back seat

All of the focus on state and national politics came with a cost. By 1961 Arthur Andersen and Company, the national accounting and auditing firm, was warning of dysfunction in MFA's management structure. "Lines of authority and reporting responsibility, as shown on the organization chart, may appear for the most part to be clearly defined. It is our impression that these lines are not all clearly defined. As an example, the division managers of grain and feed, packing, and state exchanges reportedly report to the general manager. In actual practice, they generally report direct to the management committee."

According to Ray Young, the auditors were unsettled by net savings in view of sales volume. After 1955, profitability painted "a grim picture. Per cent of savings to assets employed is regarded as unsatisfactory," the auditors warned. As Young reported in his book, "Year after year losses occurred in certain profit centers, including the poultry and egg operations at Sedalia and Shelbina, the Tire Division, and State Exchanges. By 1963, sales of MFA itself were $88 million, with savings of $820,000, of which $225,000 constituted patronage dividends paid back in cash to members. In 1964 the gain was only $408,000. The Grain Division showed a loss of $43,000, the Packing Division a loss of $100,000 and the Poultry and Egg Division a loss of $115,000. Patronage dividends paid were $544,000." Management, or more specifically, authority was the problem. No one moved without Fred Heinkel being on board. And Heinkel was gone for extended periods.

MFA's handful of astute businessmen shared the auditor's concerns. R.J. Rosier lobbied Fred Heinkel on MFA problems. In a 1965 paper, singled out in the mid-1990s by then MFA President B.L. Frew as one of the most insightful MFA white papers ever presented, Rosier identified business problems and solutions. "It is important to remember that we have some definite obligations," he wrote, "to safeguard the funds of those from whom we have accepted investments in good faith; to conduct the business of the Association in the best interest of members; to constantly expand our operations in order to progress."

> "It is important to remember that we have some definite obligations to safeguard the funds of those from whom we have accepted investments in good faith; to conduct the business of the Association in the best interest of members; to constantly expand our operations in order to progress."
>
> —R.J. Rosier

R.J. Rosier, one of MFA's best executives, outlined several strategies in 1965 to better control the business operations of the cooperative. This is his vision of a modern exchange facility. He specifically wanted only one store per county. He wanted uniformity in store and elevator design.

Proud Past, Bright Future: MFA Incorporated's First 100 Years

The problems identified by Rosier included lack of a stable management force, credit problems at the store level, bookkeeping costs, lack of system-wide control of grain inventories, outmoded facilities, overlapping trucking operations, active competition between stores to the detriment of all, non-attention to the growth of farming enterprises and the resulting necessary services, market changes undermining the existence of produce plants and general lack of good business organization.

Rosier wanted all local MFA agencies (from MFA Oil Company to MFA Milling Company to Producers Produce Company) to be merged into MFA Incorporated or at least into one corporation to assure a tighter corporate structure at the division level. He wanted to discontinue the breeding service "if it cannot be put on a profitable basis." He wanted centralized purchasing for all office supplies and fixtures as well as mailing and mimeographing across the different MFA-related companies.

MFA Incorporated and MFA Insurance Company jointly created MFA Gardens as a public service. When the two companies eventually split, the gardens became known as Shelter Gardens. A scaled-down replica of the Newcomer Schoolhouse still stands in the gardens.

He wanted one store in each county. That store's location would "conform with the potential business available after a careful survey of all possibilities." He wanted uniformity in store and elevator design. Furthermore, from a practical business standpoint, Rosier wanted one general manager per county and those managers reporting to a "General Manager of Local Operations." He would pay those individuals "commensurate with their responsibility and results of operations." From his standpoint, MFA could:

- Increase the borrowing power of the Association
- Make financing easier to establish or improve local facilities
- Develop a uniform credit policy
- Eliminate competition between locations, "thus putting our local operations on a sounder basis"
- Assure modern bookkeeping
- Centralize an increased sales force
- Eliminate duplication of trucking
- Increase the organization's buying power
- Eliminate "outside" buying without authorization
- Eliminate "the feeling among some Exchanges in the Central Cooperative that their net savings are being absorbed by those having losses"
- Streamline farm supply shipping
- Better control inventory
- Permit consolidation of bank accounts "thus effecting substantial savings on interest expense"
- Use bulk plants more effectively

"One over-all management" structure was sorely needed, Rosier reported. To consider these suggestions, Rosier requested a meeting with "F.V. Heinkel, W.W. Beckett, R.A. Young, A.D. Sappington and Herman Schulte"—all members of the management team at MFA Incorporated. Young, Sappington and Schulte were capable business executives. Beckett was general counsel. Only one man, however, could authorize such a meeting and order implementation—F.V. Heinkel. No records exist as to whether that meeting occurred. Records do show none of the suggestions was acted upon.

Richard Collins, confidante of Fred Heinkel and long-time MFA vice president of communications, maintained that Heinkel was aware of structural problems but fully engaged in the building, not managerial mode. "We weren't focused on building business structures. We were interested in helping farmers better their lives. The details, the management arrangements, the stuff Ray Young always worried about—those could wait. They weren't central to our efforts." The two positions, ideally, were not mutually exclusive. In practice, however, they too often were.

In 1965, Heinkel attended a Production Credit Association meeting in March and was impressed by one of the speakers: R.B. Tootell, governor of the Farm Credit Administration, who spoke on the future of the family farm. The No. 1 threat? Technology. "We are told that 90 percent of all the trained scientists who ever lived are alive today," said Tootell. "The implications of this are great. These men, unlocking the secrets of the universe, will discover much that has an impact on agriculture. Nevertheless, the family farm has shown a great capacity to adjust to the impact of advancing technology by using it."

Co-ops are essential, noted the speaker, but they must be efficient. Co-ops must focus on petroleum, fertilizer, chemicals, manufactured feeds and other purchased inputs. But, he noted, "The dilemma faced by cooperatives generally is that the nearer any one of them approaches its objective, the less apparent is its value." In a hand-scrawled note filed proximate to the program, Heinkel noted: "No problems—little action—progress." He jotted out a Norman Vincent Peale anecdote. "Peale knew of a community of 100,000 with no problems. Interested in joining? It's called a cemetery." Further on, he noted, "farmers are at the mercy of worms, weather and world markets." Heinkel was not unaware.

Yet he seemed not to have internalized one important lesson from William Hirth: "And yet when we hold those at the head of the Oil Department responsible for the best results, I am sure you will realize that I cannot at the same time dictate to them in such a vital matter as the matter of sites," Hirth had told an MFA member back in 1929. It was a lesson lost on Heinkel. He picked sites.

"We weren't focused on building business structures. We were interested in helping farmers better their lives."

—**Richard Collins**

An unsound strategy

Fred Heinkel had little patience for bankers, even those financial experts from the St. Louis Bank for Cooperatives who supplied millions of dollars to finance MFA's operations. Heinkel would drum his fingers on the table and look distracted during lender meetings.

Undeterred, executives of the St. Louis Bank recited a litany of MFA's financial weaknesses in the mid-1960s during a meeting they demanded. Heinkel brought in his tight group of executives and a few hand-picked board members. Harry Chelbowski, president of the St. Louis Bank, called the group's attention to five items that MFA management needed to address:

1. Run the company as a business
2. Adopt strategic planning and make sure the board is actively involved
3. Establish clear-cut policies
4. Work together as a management team
5. Communicate and cooperate

Chelbowski had come to the St. Louis Bank for Cooperatives after years in the St. Paul Bank and 30 years in Farm Credit. He knew agriculture and he knew cooperatives from stem to stern. He could not believe an organization the size and complexity of MFA elected its CEO from the membership each year. In all his years, he had never seen any other organization that would subject itself to such risk. He could list no exceptions. He and the other bankers demanded a separation of politics from the business.

Throughout Chelbowski's presentation, Heinkel continued to drum his fingers and look distracted. Suddenly, he cut to the point: "Have you ever lost a dollar on us?" he demanded. Otto Schulte, MFA's vice president of the grain division, who attended the meeting, said bank presidents and lenders irritated Heinkel. "He had his eye on accomplishment, not on paper."

It was unsound strategy. Bankers had reason to worry. So did the leaders of MFA. MFA was using too much debt to finance the business. There were too many non-productive assets. Financial planning was weak or non-existent, resulting in too many "surprises." MFA was too complex (probably the result of too many lawyers, one bank officer later quipped) for adequate control. The organization chart did not fit the business, and the board tended to do whatever Fred Heinkel recommended, whether it was financially and strategically sound or not.

No part of their presentation, the lenders would say later, seemed to connect with MFA leaders at the meeting. The lenders went back to St. Louis discouraged, but determined.

124 POLITICAL DISTRACTIONS

"Keep your expenses below your income"

Management weakness hampered operations. Many problems identified by Rosier traced their source to reporting structure. At the heart of many problems, individual store managers had no firm chain of command. Managers did have a direct line to Fred Heinkel. After all, store managers recruited board members and implemented policy at the local level. Much of that policy came from Heinkel. Stores were the backbone of the cooperative. Store managers felt free to call the president for decisions and to report local activities and local reactions to MFA decisions.

When Jack Silvey had first managed MFA Central Cooperative, there were five exchanges. By the time he moved to head insurance efforts in 1945, Central Cooperative had grown to 20 exchanges. By 1948, Otto Schulte was managing Central Cooperative. Sixteen years later when Schulte left to become manager of the grain division, Central Cooperative had 70 exchanges. His brother Herman Schulte took the reins.

This gradual migration of local cooperatives into MFA Central Cooperative underscored what Dr. Mike Cook, who occupies the Robert D. Partridge chair in cooperative leadership at the University of Missouri, would term MFA's stealth transformation from a majority federated system to a majority centralized system. There was no organized effort at MFA to implement the change, just a practice of assuming control when local cooperative members tired of managing employee turnover or fighting to maintain profitability. Letters in the MFA archives testify to the phenomenon. One simply asks for MFA to take over because members had replaced an incompetent manager with an embezzling one. They wanted to farm, not manage day-to-day affairs of a local business, especially if they had to beware of thieves.

Otto Schulte had left another asset with Central Cooperative: Gene Murphy. Murphy grew up in the MFA system. His father managed the La Grange MFA exchange. The young Murphy managed his first store (Lewistown) after a stint in World War II. "I caught the MFA bug from my dad and my granddad," Murphy explained. "I always enjoyed helping farmers. They can be hard people to work for, but they're good people."

"I caught the MFA bug from my dad and my granddad. I always enjoyed helping farmers. They can be hard people to work for, but they're good people."

—**Gene Murphy**

MFA Stores

Martinsburg

McGirk

Memphis

Meta

Mexico

Gene Murphy was 24 when Schulte hired him as manager after an interview in the cream room at the La Grange exchange. Impressed by the young man's abilities as manager, Schulte told Murphy to keep an eye out for MFA locations he'd like to manage. At the time, Marceline had a new, modern facility and in due course Murphy mentioned to Schulte that he'd had his eye on that one.

"No, I've got the place for you," Schulte told the young man. That place was Brunswick in 1955. Murphy took the job. Schulte kept his eye on Murphy and in the early 1960s asked Murphy to choose between managing the newly purchased facility at Lexington or a newly created position of district manager. It was a new, but short-lived management structure. Many of the exchanges at the time were unprofitable and managers needed help to plot the way to profitability.

"I never understood why some stores didn't make money," said Murphy. "The job intrigued me." He took the job and moved to Columbia in 1961. The district manager concept had been drawn up to better manage MFA's widespread store system. The state was divided into two sections vertically using Highway 65.

Murphy oversaw the eastern half from Iowa to Arkansas. Another individual was hired for the western portion. The structure lasted just one year before two more individuals were added. There had been no way, Murphy said, to get to each store much less understand the market and advise them. From then on, Murphy's territory ended just below St. Louis.

Schulte had moved on to manage MFA's grain division and his brother Herman began managing MFA Central Cooperative. "I considered Herman an astute businessman, but Herman didn't stay long," said Murphy. Although Herman Schulte was sharp and effective, "he wouldn't put up with a lot. He stayed maybe a year and went back to the seed division in Marshall." Interference was what Schulte couldn't tolerate. And that interference came straight from the top.

On the job, Murphy quickly learned that some exchange managers made a lot of money selling MFA insurance but did a poor job of managing their stores. Murphy didn't fault them, necessarily, although he was quick to point out that managing the store was what they'd been hired to do. What Murphy understood was that the problem was oversight.

Herman Schulte next sent Murphy to the MFA Exchange at Bernie which, although in prime agricultural territory, had lost $200,000 the prior year. Through Schulte, Murphy learned of Fred Heinkel's advice to contact a specific individual in the area and lay the groundwork for any activity. That raised Murphy's hackles and underscored the problems he'd already encountered at other facilities.

"Herman," said Murphy, "I don't know how fast I can turn it around. But if you're wanting someone to go down there as a politician, you're talking to the wrong man. I'll manage it, but I'm not a politician. And I never did call that man. He called me. That was, as far as I was concerned, the ruination of MFA at the time—pure politics." Within a year, Bernie turned in profits of $100,000.

After another year at Bernie, Murphy was back into district manager mode and was offered the job of overseeing the Columbia store. He jumped on it with the following caveat: "This is going to be a barn burner, because I'm going to make some changes out there," he told Shelton Cunningham who was vice president of operations for Central Cooperative and oversaw the district managers. "You may not like it and Mr. Heinkel, I'm sure, won't like it. But if it's going to be in my district I'm going to make some changes. There was a man up at Monroe City doing a good job—a good manager, Roger Gilbert. I conned him into coming down, told him what the circumstances were and asked him to manage Boone County Exchange." Gilbert would become a fixture at the Exchange and turn in year after year of profitability. "Shelton Cunningham was one of the best executives working at MFA but his authority was consistently undercut by Mr. Heinkel," said Murphy.

On a trip through his territory in the late-1960s, Murphy rode in the backseat. Up front, with an MFA employee driving, Fred Heinkel looked out

Political Distractions

the passenger window as the vehicle passed an MFA location. Heinkel turned to Murphy and asked, "How's this one, Gene? What are the prospects for its success?"

"Mr. Heinkel," Murphy replied, "that store is too small, too run down. Its territory is bordered by brush on one side and the other sides have good, modern MFAs too close. It will never take business from them. The area won't support it. It has lost money, it's losing money now and it will continue to lose money no matter what we do. My recommendation is we sell it."

The atmosphere in the car had been cordial, until that moment, recalled Murphy. Heinkel placed his left arm across the back of the front seat, pulled his body around until he was squarely facing Murphy. He fixed Murphy with a cold, steely stare. "MFA does not sell stores, Gene," he said icily. "MFA buys stores." Less than a decade later, the unprofitable location was closed.

Murphy would soon find himself in trouble. He preached fiscal discipline and channeled the communications flow through the chain of command. His no-nonsense approach focused on honesty and traditional business practices. The chain of command structure cut off exchange managers from their traditional straight line of communication with Fred Heinkel. Managers lost influence. Key managers with long ties to Heinkel protested and called Heinkel with complaints. Heinkel took those calls and advised the managers they would be exempt.

As another influential MFA business executive, David Jobe, would complain in later years, a major management frustration at the time was that many exchange managers had as much, if not more, influence with Fred Heinkel than did the vice presidents who reported to him. "It made policy decisions unenforceable. And policy adherence in a business structure is critical. Fiscal discipline falls apart without it," said Jobe.

Murphy, ignoring the politics at his peril, tightened control over stores, visiting each manager and setting expectations. It wasn't long before he learned the consequences.

Cunningham called all the district managers into the home office and said, "We've eliminated your positions." The new management positions had put a layer between Fred Heinkel and store managers, his direct source of information in the countryside. Heinkel found the situation intolerable. Cunningham, himself, would soon be replaced.

Murphy would land on his feet. He was hired initially to manage a Farmland affiliate at Fayette, but soon found himself in charge of Ray-Carroll County Grain Growers. When Murphy came aboard, the cooperative was suffering from financial setbacks, a situation he promptly turned around. By the time he retired, Ray-Carroll had an excess of $8 million in working capital.

"Early on in my career R.J. Rosier had stopped by my store. He gave me good advice," said Murphy. "'Gene, there's only one way to make money,' Rosier said. 'Keep your expenses below your revenues.' I never forgot that."

Beware the old lion

Back in 1964, Jack Silvey, who had proved himself a stellar business executive, left the company. Ostensibly, the reason was Silvey had fired an MFA department head who was popular with both the board and Heinkel. Silvey had successfully managed Central Cooperative, begun the farm supply division and been chosen to lead MFA Insurance. He'd been highly effective in all roles. The new company grew rapidly, and, under Silvey's oversight, had expanded into multiple states. But Fred Heinkel's corporate board members weren't Silvey fans.

Silvey's focus was bottom line, not placating people. Furthermore, Heinkel suspected Silvey of ambition. So when Silvey dismissed the employee, Heinkel confronted him. The result? Silvey resigned. He went on to form his own insurance company and died decades later a wealthy man.

Then there was William Beckett.

Beckett was not easily intimidated. As general counsel for MFA since 1957, he'd fought many courtroom battles and faced down dozens of opponents. More indicative still, Beckett was a decorated WWII combat veteran. He'd served in England, Algeria, Tunisia, Italy, Germany and France in six campaigns. He had three Purple Hearts and the scars to prove it: he'd been stitched by a machine gun, blown up by a tank and run over by a half-track. But in 1965 when a St. Louis reporter featured Beckett as the quiet, unassuming but second-most powerful man at MFA who would likely succeed Fred Heinkel, Beckett hastened to assure Heinkel the reporter had taken liberties with the facts and made unwarranted assumptions.

Beckett harbored no such ambitions, he promptly explained by letter to Heinkel, "I want to say once again, as I have said many times in the past, I have no ambitions in this direction, either before or after your retirement. I am concerned only with the success of the MFA, and I would like to be regarded only as a competent and interested employee."

Beckett knew the score.

Fred Heinkel was 68 years old. Rumors regularly surfaced as to when he would retire. That August at the convention, friends and members, after unanimously electing him yet again, had presented Heinkel and his wife with a certificate of appreciation and money for a trip around the world. Heinkel, fearful the trip was a hint to retire and that his absence would give room for maneuvering, refused to go. He harbored no thoughts of retirement and wasn't about to step out of the arena. In fact, he dealt harshly with those who looked as if they wanted his job.

Despite being on the ticket as vice president of MFA with Fred Heinkel in 1967 and having full responsibility for MFA operating divisions and the related staff of the controller's office, central accounting and data processing, Beckett quietly left MFA in fall of 1968.

Unfortunately, Heinkel was beginning a pattern that would foreshadow future events.

Make no mistake, Fred Heinkel was a great man who did great things for farmers. He tenaciously held his position because he fully believed, with ample evidence, that his control was necessary to continue building MFA's (and the farmers') future. As a man who had built MFA into a national powerhouse, Heinkel had no equal. Ever the visionary, he had miles to go before he slept.

As testament, Heinkel was soon organizing efforts to develop MFA Foundation, a nonprofit Missouri corporation dedicated to funding scholarships for rural youth. Established by a donation of $28,000 in 1958, the foundation languished until Fred Heinkel took an active interest in the early 1960s and oversaw its investment and then development into a foundation that provided, at first, $200 scholarships to worthy students. The foundation provided $100 of the scholarship; the local participating MFA agency footed the other $100. Over the decades, the foundation has provided more than $15 million to thousands of young people.

Speaking at Farmers Produce Exchange #139 in Lebanon in 1974, Heinkel outlined four stages of MFA's development.

> The first step was the Farm Clubs placing orders for shipments of carloads…It was easy to see and measure the savings that could be made by pooling purchasing power…Then in the early '20s the next step came along—the organization of Farmers Exchanges, and in the beginning those savings were easy to see, and they were considerable. Soon thereafter the third step was undertaken by the organization of processing plants to market eggs, poultry, cream, and milk and to manufacture feed. … The next step in the evolution of business activities of MFA was our joining together with other regional cooperatives to own and operate manufacturing facilities…

There would be one more step in MFA's evolution: professional management. And one impediment: Fred Heinkel.

A concession to management

By 1968, Fred Heinkel relented and made concessions to the bankers. Heinkel turned to Ray Young, a trusted, adroit executive, who'd led MFA Oil Company since Hirth's days. The corporate board confirmed Young as MFA's executive vice president in November of 1968, as well as a board

> *MFA's total assets over the decade of the 1960s had grown from $22 million to $60 million and sales had grown from $53 million to $104 million.*

member of several interregionals. Young would also continue in his role as executive vice president and general manager of MFA Oil Company.

Ray Young was all about management, balance sheets and prudent operation. Young was respected by Fred Heinkel and MFA vice presidents, but with qualification. Up to 1968, Young would attend portions of Fred Heinkel's staff meetings, deliver his opinions on issues and leave to attend other pressing issues. After his exit, Heinkel would disparagingly say, "Well, we've heard the report of the American Management Association. Now how should we proceed?" The attitude undercut Young's effectiveness with his peers.

Undaunted, Young assumed full responsibility for finance, grain, management services, general counsel and employee relations. And he took charge forcefully. One of his first official acts was to initiate plans to sell American Press, which had accumulated a loss of $351,000. Recall that William Hirth owned a printing press that he used to print The Missouri Farmer. With the purchase of The Missouri Farmer in 1940, MFA also acquired Hirth's printing operations. Over the course of almost 30 years, the printing operation had morphed into a large commercial press that MFA christened American Press.

What MFA officials did not do, however, was run the company profitably. Young made almost immediate arrangements to sell 49 percent of American Press to a professional printer. In so doing, he kicked a hornet nest. Management rebelled. "I felt that perhaps I had been too optimistic about the freedom I would have in running the business affairs of MFA," he recalled in his book. Nevertheless, Young argued his case and the deal remained. By 1971, Young would be able to sell the remaining 51 percent of capital stock of American Press for current net worth of $459,000.

For 1968 MFA would post a $400,000 loss. By the end of the following year, though, under Young's able guidance, MFA was profitable to the tune of $100,000, still a sum far below acceptable metrics of return on assets. But it was a start. MFA's total assets over the decade of the 1960s had grown from $22 million to $60 million and sales had grown from $53 million to $104 million. Clearly, lack of effective management was the problem.

And the bankers, feeling Young's actions were insufficient, took action.

The St. Louis Bank for Cooperatives classified MFA's operating loans in 1969. The term classified indicates a problem loan. Red flags go up on banker radar. Of the bank's 300 loans, only perhaps as many as five would rank this attention. Bankers served notice that MFA's position was caused by heavy liability in relation to member equity, low profitability and low working capital. The bank, said Harry Chelbowski, would accept some of the blame because bank officials had not been firm enough.

Ray Young took the matter to heart. He immediately addressed management structure. He hired a consultant, controlled expenditures, reduced operating costs through consolidation and developed wage controls. He then miscalculated. He brought in an executive from Standard Oil, an individual Young described as "highly competent," to head supply and marketing. The executive, in turn, brought in another individual from Standard Oil. The two set up a reporting structure that included field men, similar to what had occurred in retail but had ended badly. This time the result was no different. By 1971, Fred Heinkel approached Young and forcefully demanded both individuals be fired and the structure dismantled.

By 1970, at Young's urging, MFA Central Cooperative and existing state exchanges would be rolled up into what would be named the exchange division. Since Central Cooperative was a separate cooperative, a membership vote had been required. From a professional management perspective, consider the process: To streamline management functions, to make a decision on how to best operate an essential portion of the company that, by anyone's estimation, needed restructuring, MFA executives had to call for a vote of the membership to enact the change.

After an affirmative vote, MFA executives could more effectively streamline and manage the division. But that effectiveness would be several years in coming. For its first full year of operations,

▲▲ MFA Central Cooperative began as a way to keep financially shaky exchanges afloat. Many early exchanges ran into problems with management but were located in productive areas. Often, over the years, farmers and ranchers on local boards grew tired of hiring and supervising managers. MFA would step in and run the operations profitably.

MFA Stores

Moberly

Montrose

Mountain Grove

Neosho

New Haven

the exchange division posted a loss of $2.4 million. The manager was replaced by a young engineer who had been brought in from FS Services (now Growmark). His name was B.L. Frew, and he would become both a great MFA asset and a catalyst in the downfall of Fred V. Heinkel.

By 1972, MFA operations were showing marked improvement. MFA added millions of bushels of grain storage and tightened control over many previously wayward divisions. The cooperative ended the year with net earnings of $2.7 million. Under Frew's leadership, the exchange division added $2 million to the bottom line.

The 1970s also saw MFA increase investment in interregional cooperatives to both good and bad effect. To the good, the interregionals showed substantial earnings and helped increase MFA's (and customers') supplies of products. To the bad, the influx of patronage from the interregionals led to reliance on that revenue stream at the expense of profit on MFA's own operations. After all, if an interregional would return $10 million in patronage, why ruthlessly manage operations? Fred Heinkel would use those figures to counter Young's techniques.

MFA invested in Cooperative Farm Chemicals Association and Mississippi Chemical Corporation for crop products and CF Industries for plant foods. MFA invested in Agri-Trans Corporation, an interregional organized to haul grain down the Mississippi and plant foods up the river. It included a fleet of riverboats and barges.

Fred Heinkel and Otto Schulte would also direct MFA's participation with other cooperatives in building the Farmers Export Company, a grain interregional near New Orleans. The elevator was built in 1969. Heinkel would serve as chairman of the board for several years.

By 1975, the lenders would "declassify" MFA's operating loans. Still, they viewed MFA's structure as weak and directed their officers to key sharp eyes on MFA's activities. The same officers from the St. Louis Bank again met with Heinkel, his selected executives and board members. They repeated their initial presentation. Heinkel repeated his performance. The bank officials stood firm. From their perspective, MFA's financial planning to date had only been a tentative step. They wanted MFA leaders to state specifically where the cooperative was headed and how they were planning to get there. Despite Young's efforts, the bankers said, MFA was just reacting to whatever happened in a given month or year. The problem in the financial arena continued to be the president.

MFA had a long relationship with St. Louis Bank for Cooperatives. The bank officials helped MFA work through financial problems in the late 1970s and early 1980s.

Buying time

As Fred Heinkel approached his 80th birthday in 1977, he would lean more and more on Clell Carpenter and familiar political issues, ignoring the pesky bankers. Young and other executives, including Bud Frew, would complain of MFA decisions continuing to be driven by political rather than strategic considerations. If a corporate board member with clout (important to keeping Heinkel in power) wanted a bulk plant in his area, the plant would be commissioned regardless of its effect on business operations. Fred Heinkel still led and made final decisions, overruling Young.

For years, in fact since Hirth's reign, conventions had always drawn supporters from across the political realm. Recall, for instance, Hirth's frequent exhortations that "I am not a partisan," a refrain he would repeat again and again throughout his life to explain his political inconsistency. Hirth supported members of both parties but most frequently found himself campaigning for Democrats despite his fast friendship with Republican senators, representatives and governors. From Hirth's perspective, he was for agriculture, first and foremost. Party affiliations were superfluous. What mattered to Hirth were "fearless men." In the 1932 campaign covered in previous pages, Hirth would support a Republican for state senator while campaigning for Roosevelt.

For several years, Heinkel at least attempted to follow that principle. As late as the 1966 convention, Heinkel booked both U.S. Senator Bob Dole of Kansas and U.S. Vice President Hubert Humphrey, one of Heinkel's friends.

By the 1970s, however, Heinkel was clearly a staunch Democrat. The 1978 MFA annual convention was a virtual festival of Democrats, drawing U.S. President Jimmy Carter, U.S. Secretary of Agriculture Bob Bergland, U.S. Senator Herman Talmadge from Georgia, U.S. Senators Tom Eagleton and Jack Danforth of Missouri, U.S. Congressmen Richard Gephardt, Ike Skelton, Tom Coleman, Gene Taylor, Richard Ichord and Bill Burlison, as well as Missouri Governor Joe Teasdale. Danforth, Coleman and Taylor were the only Republicans.

U.S. President Jimmy Carter attended the 1978 MFA annual convention at the bequest of Fred Heinkel. MFA, at the time, was strongly identified with the Democratic Party.

Fred Heinkel was at the height of his business and political power in 1960. He would serve another 19 years before being defeated for the presidency of MFA for the first time. Born in 1897, he was almost 82.

The mid-to-late 1970s also saw some of MFA's best years of profitability, bolstering Heinkel's claim of effectiveness. But for executives like Young (and bankers) who looked closely at financials, a weakness was glaringly obvious. The majority of MFA's profitability came in the form of patronage checks from interregional cooperatives of which only half was payable in cash, the rest in equity.

At the 1974 Lebanon meeting where Heinkel outlined the steps in MFA's evolution, Heinkel would report on

> the greatest volume of business and the highest earnings that we have ever had for the year which will end on August 31. Total sales will be in the neighborhood of $450 million, with net savings of $26 million. Almost half of those earnings are a result of investments that were made by Missouri farmers in basic fertilizer manufacturing facilities and in grain marketing organizations. The board of directors of MFA has voted to pay back 50 percent of net savings for the current year, and, in addition, will go back and pick up $2.5 million of the oldest equities. Out of these old equities, your Lebanon Exchange will receive approximately $15,000 in cash, and in addition 50 percent of the current year's earnings on business done with the MFA will also be paid in cash.

To the astute executive focused on metrics, MFA's balance sheet would show, in the instance above, $11.9 million in the total patronage received. But of that sum MFA received just $7.3 million of that patronage in cash. The rest was equity. In turn, Heinkel mandated MFA pay out $14.7 million patronage in cash to customers. The result? A real cash debt resulting in liquidity decline as MFA burned through cash in the 1970s. MFA was not retaining enough capital.

By 1977, working capital as a percent of assets was down to 4.45 percent. Years prior, in his prime, Fred Heinkel had vociferously argued against paying the maximum cash patronage. In fact, in years past, by retaining earnings, Heinkel had enough cash on hand to start an insurance company and to invest in a plant foods interregional. Now by paying large patronage checks, Heinkel was buying time.

In 1979 when Fred Heinkel was just shy of 82, time ran out.

PROUD PAST, BRIGHT FUTURE: MFA INCORPORATED'S FIRST 100 YEARS

Bud Frew came to MFA in 1970 as director of operations of the exchange division. Within a short time, he was promoted to vice president of the division. Frew was a no-nonsense businessman and worked to bring the division into the black.

★★ Chapter 7 ★★
Transfer of Power

Translated from the original Anglo-Saxon

B.L. (Bud) Frew didn't suffer reprimands well. But this time, he kept silent and listened. He had to; he was in total agreement. After all, a handful of managers in 1977 (far too many from Frew's perspective) had "violated company policy by buying futures, that had nothing to do with merchandising, on grain"—and lost. Lost big. Ray Young, on his personal letterhead, put it in writing, called the actions speculation resulting in "abnormal losses" by exchange division personnel.

Frew wrote new procedures and controls. And in his blunt, impersonal way, he made consequences known for future infractions in no uncertain terms. Frew was MFA's vice president of the exchange division. From here on out, he'd be keeping an even closer eye on managers and their daily position reports detailing grain activity.

Frew had come to MFA in 1970 as director of operations of the exchange division. He would quickly move up to vice president of the division. A mechanical engineer, he had worked for 10 years at FS Services (today's Growmark). Frew was a no-nonsense businessman. He kept both eyes on efficiency and business statistics. In his second year at MFA, Frew drew up a list of 20 exchanges that individually were losing approximately $10,000 a year each and every year. He wanted them closed or consolidated. As his report circulated, political considerations accumulated. Some closed, others remained Frew's problem.

MFA's financials had been on Frew's mind. From his talks with Young, Frew was well aware of banker concerns. Those concerns were his as well. He wasn't satisfied with operations. From his position of vice president of the exchange division, Frew looked high and low for ways to keep MFA in the market, price-wise. High overhead costs and intercompany allocations outside his control translated into less margin or higher price. Either or. And his margins were razor thin.

One other aspect straddling MFA's weak balance sheet was high investment in interregional cooperatives. On the one hand, interregionals provided needed product in tight times. On the other, interregional profits tended to consist of a high percentage of paper and locked MFA into a market that could be out of position. That made moving product difficult.

At MFA's annual managers meeting in 1978, Fred Heinkel, followed by several wholesale vice presidents, stood in front of the assembled crowd and scolded those attending. Retail was not pulling its weight, he said. Other parts of the company, like MFA Insurance, were having good years. Retail's performance was indicative of lack of initiative. Retail was squandering farmer loyalty the organization had built over decades. The speech was accepted silently, but exchange managers seethed. None more than Frew. "How in the world could the insurance market possibly compare with retail operations in agriculture?" Frew wondered. Especially a retail market in which Heinkel's decisions were law, not Frew's.

On the program himself the next day, Frew took the podium purposefully. Retail shared only a small part of the blame, Frew told the crowd. MFA's antiquated price structures hampered product movement. MFA's retail managers were the backbone of the organization, Frew thundered. The audience seconded that motion with a standing ovation. Farmer loyalty, by Frew's calculations, wasn't being squandered by the retail division. Frew knew loyalty was, in large part, a function of economics. Loyalty had never meant blind faith. High performance draws loyalty, Frew said to anyone who would listen. An allocated overhead from non-productive assets existed in the labyrinth of MFA's structure. That structure included chronically unprofitable locations he was powerless to close. Frew and his exchange managers had their hands full moving product. Margins were all important—whether through inputs or grain.

So when in late 1978, auditors told Frew of a manager misstating his grain daily position reports to cover losses and keeping open positions in direct violation of company policy, Frew pushed for further audit reports. Those reports confirmed the initial auditors' conclusions. Frew fired the manager of the company-owned exchange.

And all hell broke loose.

At the time of the manager's firing, Fred Heinkel was wintering in Arizona. As had been his custom during his late 70s and early 80s, he and his wife journeyed to Arizona for much of the winter, spending up to three months in the warmer climate. Correspondence back and forth with MFA's corporate office occurred mostly by mail, sometimes by phone.

Within a short time of the manager's dismissal, Heinkel received calls from supporters of the manager and promised he'd look into the situation when he returned from Arizona. Meanwhile, politics kicked into high gear. Clell Carpenter began updating Heinkel on the political ramifications of the situation, and within a short while began lobbying for the manager's reinstatement.

In conjunction with the phone calls and personal visits, there were protests, community meetings, and even a blockade of the exchange. Remember, this was the 1970s, even in the conservative Missouri heartland.

When Heinkel returned to Columbia, he authorized an outside, independent audit. The results? Yet more confirmation of the original findings. But the audit found no criminal activity. USDA even weighed in in support of the dismissal, finding "serious discrepancies."

Heinkel met with Carpenter and Young and asked for opinions. Al Hoffman, MFA's general counsel, had outlined four scenarios: reinstate the manager; refuse to reinstate him; transfer him to a different job with no oversight of exchanges; or sell the facility to the protesting members.

Hoffman made no recommendations. He did not, however, refute the original audit findings. They stood. He did point out the drawback of the fourth alternative. To acquiesce in that manner would set a dangerous precedent of "a rule by force rather than a rule of reason in operating the company," the attorney wrote.

Young was adamant they should refuse to reinstate the manager. He vociferously argued the necessity of maintaining the chain of command. To maintain fiduciary responsibility, argued Young, Heinkel should support the decisions of those to whom the upper management had entrusted the operation of the business. He said flat out that Frew would resign if his authority to operate the exchange division was undermined. How could any manager control employees who knew the manager's decisions weren't final? Frew was instrumental to Young's organizational structure. Frew understood the financial straits and worked hard with Young to improve the cooperative's fiscal condition. Frew was a shining star.

First thing the morning after that meeting, Young stopped by Frew's office and poked his head in. "You had breakfast yet, Bud?" he asked.

Frew said, "If you're here to tell me what I think you're here to tell me, I don't need breakfast to give you my answer." Young deftly swept aside the response and convinced Frew to join him at a local restaurant. Using all his finesse, Young tried to soften the effects of Heinkel's decision, tried to keep Frew from resigning—but to no avail. As Young would later note, Frew's response would have to be translated from the original Anglo-Saxon. Frew quit.

MFA was in his blood.

Bill Stouffer still farms at Napton, outside of Marshall, Mo. In the late 1970s, Stouffer grew disenchanted with MFA policies under Fred Heinkel's leadership. He made an appointment with Heinkel and traveled to Columbia to talk with him. Stouffer wound up running for a seat on the board—and winning.

A historic friendship

To this day, former Missouri State Senator Bill Stouffer bleeds MFA red, white and blue. His father was an MFA delegate; his grandfather a member. Stouffer grew up knowing MFA was built by farmers for farmers. In the early 1960s, he received an MFA scholarship to attend the Missouri Institute of Cooperatives. He received the first college scholarship offered by the Marshall MFA Exchange. His first off-farm job was at the MFA Exchange in Columbia, part-time while he attended the University of Missouri. MFA was in his blood.

Back home in the 1970s after finishing his degree in ag economics, Stouffer chaffed each time he visited the exchange for his farm inputs or to deliver his grain. That irritation blossomed into full-blown anger at a seed meeting at Marshall put on by a salesman of the MFA seed division.

The salesman compared two of MFA's best hybrids to "a good Pioneer and a real Pioneer dog," said Stouffer. The salesman said the comparison proved MFA's average was better. "I knew all four seeds. I went straight to the manager and asked why my cooperative was lying to me."

Bob Ferguson was the man Stouffer confronted. Ferguson managed the Marshall exchange at the time. Years later he would rise to regional manager. Without saying much of anything, Ferguson simply directed Stouffer to another individual who had similar complaints. After talking to that individual, who pointed Stouffer to yet another dissatisfied MFA member, Stouffer found a group at St. Joe, a group in the southwest, a group around Boonville; Stouffer would find brush fires pretty well all over the state, he said. Fred Heinkel's strategy, said Stouffer, was to keep all these brush fires contained locally.

Eric Thompson was director of employee relations at MFA when he decided to run against Fred Heinkel for president of MFA. Thompson was 36. Fred Heinkel, in 1940, had been 43.

Not one to let matters drop, Stouffer went straight to the top. "I've always been one to work within the system," he said. "I don't throw bombs from outside." He made an appointment in the spring of 1979 with Fred Heinkel in MFA's headquarters to talk about his concerns. Heinkel welcomed the young man into his office and listened to his complaints.

"He brought in the head of the division," said Stouffer, "and basically I was told I didn't know what I was talking about. Meeting Heinkel was like meeting a prize-fighter. He kept dodging and weaving." The individual in charge of the seed division told Stouffer the seed comparison was "just the way you do business." Stouffer turned away from the division head and said, "Mr. Heinkel, if this is going to be Heinkel Feed and Grain, I'll go home and I won't say another word. But if it's truly Missouri Farmers Association, that's not the way my cooperative is supposed to do business. I want you to be straight with me." He left without a commitment.

Upset, but undeterred, Stouffer told his wife he wanted to talk to the individual whose firing as vice president of the exchange division had caused such uproar. "I'm going to call this devil up," Stouffer told his wife, "and see what he looks like."

Bill Stouffer called Bud Frew. A life-long and historic friendship began.

Stouffer arranged to meet Frew at a local restaurant. Stouffer explained his frustrations and asked for Frew's insight. "Frew sat right there and drew out on a napkin what MFA ought to look like," said Stouffer. "I don't know how long we talked. It was probably an hour and a half. But from that time on, it was very evident that the company's direction had to change."

Bill Stouffer found a mission.

He went home and organized a meeting to be held on his farm outside Napton. He invited all the

Eric Thompson sought out Bill Stouffer to determine who he would support to run against Fred Heinkel. Shortly thereafter, Thompson approached Ray Young, executive vice president, to see if he would stay on if Thompson won. Young refused to take sides.

farmers he'd visited with who shared his concerns. Two dozen from all over the state showed up. After hours of conversations, both at that meeting and at meetings that followed, the group decided on a course of action. They would run themselves or support those running for seats on the MFA corporate board of directors as long as those running reflected their desire for change. On the ballot in 1979 were Carlton Spencer, Faucett, Mo.; David Hortenstine, Brookfield, Mo.; L.E. Manson, Brunswick, Mo.; Everett Billings, Green Ridge, Mo.; Adrian Murray, Ash Grove, Mo.; Bill Stouffer, Napton, Mo.; and William Umbarger, Fairfax, Mo. Seven men determined to fix their cooperative, eerily reminiscent of another meeting of seven farmers 65 years earlier.

It wasn't MFA employees rebelling, Stouffer insisted; it wasn't a disgruntled Frew pushing for change in leadership at MFA as payback. "It was member-driven," he said. Period. The millions of dollars MFA spent on lobbying and governmental issues, he said, were showing up along with costs from an outdated organizational structure. Farmers were paying the bill. It was the farmers' money MFA was mismanaging, Stouffer said. That had to stop. "We had a 1950s distribution system" supporting a 1979 agriculture, he said.

Stouffer's next step was to approach Ray Young and ask if he'd run against Fred Heinkel. Everyone at the time knew Young's strength as a businessman and knew he was the glue holding the business together. It wasn't the first time Young had been asked. Several influential MFA leaders had posed the same question before (some quite a few years before), Young recalled. In fact, over the course of the 1970s, Young had been approached three separate times. As Young reported in his book, he refused each time, saying, "I did not want to run against Heinkel, with whom I had worked almost 40 years." Nevertheless, Young would not oppose efforts to unseat Heinkel. Young knew first-hand the 81-year-old Fred Heinkel should retire—should, in fact, have retired years ago.

In the midst of all this, Eric Thompson was fuming on the sidelines. Thompson was director of employee relations at MFA. He'd heard farmer rumblings and manager grumblings. Thompson had done some of both, himself. "You have to understand," Thompson said, "the countryside was riled up." And so were exchange managers. Frew had been very popular. Thompson, as well, considered Heinkel's undercutting Frew's authority as a kick in the face to anyone paying attention. Thompson paid attention. He found out Bill Stouffer had contacted Young. So Thompson sought out Stouffer.

"We were looking for a candidate," said Stouffer. "By the first of July, I was meeting with Eric."

Thompson then approached Young at his home to find out the depths of his loyalties. Specifically, Thompson wanted to know whether Young would stay on if Thompson were to unseat Fred Heinkel. Young said he knew enough about political moves to keep things vague. Young would only say this: "I'm an organization man. I'm in this for the good of MFA. I'm not here to serve individuals, either you or Heinkel. That's all I'll say."

History and happenstance

Eric Thompson suffered no indifference. Neither did he inspire it. Friend or foe. Hero or villain. There were no shades of gray softening Thompson. An aircraft wave commander responsible for 18 B-52s and more than 100 personnel from 1969 to 1974, Thompson led men into battle. With more than 2,000 hours, 188 combat missions over North Vietnam and a Distinguished Flying Cross medal to his credit, Thompson had leadership down cold. In battle, he matured quickly.

By turns inspirational and polarizing, Thompson loved and hated. He judged character quickly, rarely reassessed and never retreated. With a master's degree in ag economics earned before his Air Force stint, he joined MFA right out of the military in 1974.

History and happenstance picked exactly the right personality to challenge the aged lion. An aircraft commander responsible for men's lives, Thompson was custom-made to step into the arena, qualified or not. Quick on his feet, cocky (some would say arrogant), unafraid of challenge and battle-tested, Thompson, after his conversation with Stouffer, considered his options but didn't hesitate long. Thompson understood Young would not step up. Someone needed to.

On July 17, 1979 (his 36th birthday), Thompson holed up in a hotel room and planned strategy. His first decision: this could not be preannounced. MFA's convention was Monday, Aug. 6. Thompson thought in military terms. "It had to be a blitz campaign," he said. "We've got to hit them so fast and so hard that there's no way to respond."

Thompson printed a four-page mailing: a two-page resume and a two-page delegate letter. Next, he mailed a letter to Fred Heinkel on July 27, a Friday, and informed him of Thompson's "candidacy for the office of President of the Missouri Farmers Association. It is with no disrespect for you that I make this announcement; however, because of the principle [misspelling in original] issues involved in this election, I have no other personal choice but to be a candidate. This is the year for change if MFA is to survive the 1980's." He ended with, "I personally was not brought up to sit idly by and complain while doing nothing. Therefore, I intend not only to be a candidate but to win."

Thompson also composed and sent two other letters that day: one to Senator Howard Baker (a Republican) of Tennessee and one to Missouri Senator Tom Eagleton (a Democrat), both of whom would be speaking at the upcoming convention. The essence of those letters? Thompson wanted both men to realize this was a contested election. They were not there to campaign for Fred Heinkel.

Thompson simultaneously sent his four-page missive to the delegates. Thompson's letter declared Ray Young should be retained as vice president, a position for which Young was currently running along with Fred Heinkel as president. Thompson also pledged something novel in terms of MFA's long-standing practice: "As I weighed my personal qualifications prior to making the decision to be your candidate, I recognized that the Incumbent was in his early 40's when he took office. Therefore, the age difference is not that significant; the difference is the fact that I have no intent or desire to serve longer than 6 years." Thompson also proposed to limit the board members to terms of six years. "If I picked six years for the board," he said years later in explanation, "how could I pick less for me?"

> *"We've got to hit them so fast and so hard that there's no way to respond."*
> —**Eric Thompson**

TRANSFER OF POWER

U.S. Senators Howard Baker (center) and Thomas Eagleton (right) spoke at the 1979 MFA convention. Thompson had written both men to inform them this would be a contested election. Thompson didn't want either man campaigning for Fred Heinkel. They would comply.

MFA Stores

Nind

Odessa

Osgood

Ozark

Pilot Grove

On July 30, a Monday, Young was in Chicago attending a board meeting of CF Industries. He received a panicked call from MFA headquarters ordering him to return at once. Fred Heinkel was sending the company plane to meet him. Heinkel wanted three things, and he wanted them right now. First, Young should renounce Thompson in a letter to the delegates, deny he and Thompson were on the ticket together, and, most importantly, offer his support for Fred Heinkel as president.

Young complied, writing the letter on the plane and mailing it the next day. In the interim, Bud Frew, who had to this point remained in the background, joined the fray. In a July 29 letter to MFA exchange managers, Frew threw his support behind Thompson. "He and I and many others feel very strongly that it is NOW or NEVER [emphasis in original]! If those of us who are genuinely concerned about the future of the Organization are to make the changes necessary to keep MFA meaningful for the Farmer Member, we must act responsibly THIS YEAR [emphasis in original]!" Frew urged the managers to encourage delegates, board members, alternates and members to stand with Thompson.

Young minced no words in his letter: "The position of President of an organization of 175,000 members with sales approaching one billion dollars, assets of over $300 million, requires more experience, in my judgment, than four years of personnel work." He stated bank officials would not "look with favor on a relative newcomer, with no experience in cooperative leadership."

The very next day, Thompson unleashed the next salvo of his blitz: another delegate letter. And he focused on bankers as well. "The Bank for Coops is concerned. However, their concern is NOT [emphasis in original] centered around me as a 'newcomer'; it is centered on our available working capitol [misspelling in original], our debt to equity ratio and our net worth. The Bank is trying to strengthen MFA's financial structure by making the loan covenants more restrictive. Their concern is basic management."

Spooked and outfoxed, Heinkel wrote delegates Aug. 1 that he would employ some of the very changes many of the new board candidates were proposing. He would also make a big concession: "For personal reasons, I do not feel that I should continue to carry the full responsibility of the office of President indefinitely. I have informed the Board that I would be willing to serve for one year in a position as Chairman of the Board of Missouri Farmers Association, Inc. and President of Midcontinent Farmers Association in order to ensure a smooth transition for the future of your MFA." He mentioned Ray Young as his likely successor.

Thompson hit hard once again the following day with a delegate letter questioning the timing and sincerity of Heinkel's latest concession. "What are we to believe about the Incumbent's future plans?" wrote Thompson. "He has implied but not promised that he would serve only one more year. He has made this statement in the past—for instance, 1969. Prior to my candidacy, the option of serving one more year was suggested; yet, he refused. Then one month later at the July State Board Meeting, he stated that he would announce his future plans in 2 years. Now in the face of defeat, he asks to serve yet another year. Can we believe it?"

And Thompson wasn't finished: "There have been many rumors started by the Incumbent's allies. They have stated that if I become President there would be wholesale firing, bonds would be in danger and the Bank will withdraw their lines of credit. When I become President, operations will be carefully and professionally evaluated. Changes will occur but only after careful study." Thompson, self-assured, included the phone number for the St. Louis Bank for Cooperatives and urged doubters to call. As he would find out after the election, the bankers were flat out furious at being drawn into a customer's election battles.

And one final salvo in the blitz, this one Aug. 3: "I believe with all my heart that if you do not choose to change leadership now, no change will ever occur.… Again, I am totally confident that Mr. Young will serve as our VP when I am elected. Because of the confidence and respect we hold for him, we know that he will serve at your request as my VP. His abilities and experience will be invaluable during the transition period…Please help me make the changes necessary for MFA to become a viable organization of the future."

A stunned silence

When Eric Thompson walked into the Hearnes Center on the campus of the University of Missouri, he walked into a crowd of more than 6,000 MFA members. Thompson would later recall the grit it took to make that walk confidently. Strategist to the end, he planned his mission down to the tiniest details. He'd mapped out the state, identified counties and delegates per county. He had a good idea of who would be voting for whom.

He had also planned his convention environment. "I remember thinking I want to concentrate on sitting with family and looking at friends," he recalled years later. "You could smell the hostility in the air in some places. I had my brother and his wife behind me. Peg [his wife] by me. Cullen Cline [his attorney] by me."

The convention program was extensive. It began that Monday morning with a call to order at 9:45 a.m. by President Fred Heinkel. D.T. Weekly, an MFA director, offered an invocation. Next the mayor pro tem of Columbia welcomed those attending. Robert Maupin, secretary of MFA (and a man who would rise to president of Shelter Insurance Companies), presented the 1978 minutes. Young followed with a business report. Sen. Eagleton addressed the crowd. Immediately following his speech, Heinkel presented the senator with MFA's "extra mile" award in appreciation of his service, leadership and accomplishments as chairman of the subcommittee on agriculture and "his successful efforts to assure that midcontinent farmers receive adequate fuels for their farm production needs."

President Heinkel next addressed the audience. That speech is not in the corporate files. Next, the credentials committee was presented and the convention broke for a picnic lunch supplied by MFA Insurance Companies.

At 12:45, delegates of MFA met in assigned locations to select nominees for directors. By 1:15 Heinkel called the meeting to order with Howard Lang, president of MFA Insurance Companies, presiding. Lang asked for a moment

▲ District number signs were used during MFA conventions to identify the districts where delegates would caucus.

▼ This is Ray Young's convention brochure. Someone had given him a heads-up on the election results from early districts.

Ray,
We have reports from District 1, 2, 3 & 4 and Thompson is leading big in all four. No reports yet from other five districts.

of silence for A.D. Sappington, his predecessor who had died just months before. Sen. Baker next addressed the group. Next Earl Manning, an editor of Progressive Farmer magazine, presented Heinkel with the magazine's national "Man of the Year in Agriculture Award." Heinkel then presented MFA's Award for Distinguished Service to Agriculture to Gene Taylor, a Republican U.S. Representative from Missouri.

After a resolutions committee report by Walter Leuhrman, both candidates for president were given five minutes to speak. There are no copies of Heinkel's speech in the files. Both candidates were to speak only from the floor, not upon the podium. Thompson led off. As Thompson rose to speak, his supporters cheered wildly and loudly. But when Thompson spoke, to Young's horror (as Young would later explain, "I couldn't believe the s.o.b. did it."), Thompson said if the membership were to elect him president, "I was personally assured in a meeting with Ray Young last Wednesday morning that he would serve as my vice president."

Young was in the crowd. In fact, he was sitting right beside Heinkel. "Dammit, that's not what I've told him," Young said to Heinkel. Heinkel turned toward Young and glared. "Then go up there and tell it different," Heinkel demanded. Young hurried down to the podium, but the rules agreed to by both candidates before the election would not allow response.

As the assembled throng held its collective breath, the votes were tallied: Thompson, 1,088; Heinkel, 615. A stunned silence hung in the air. Even the most vocal of Thompson's supporters stood mute. Observers would later comment that no one for the moment could believe what actually happened. Fred Heinkel, who had won the vote 38 times, stood defeated. All seven of the new board candidates were successful. Nothing less than total victory for the "renegades."

Still, the audience would come to its feet with thunderous applause in recognition of Fred V. Heinkel's legacy of accomplishment.

Years later, after Heinkel's death, Young would sadly note: "Eventually, had he not been unseated, he would have run MFA into the ground."

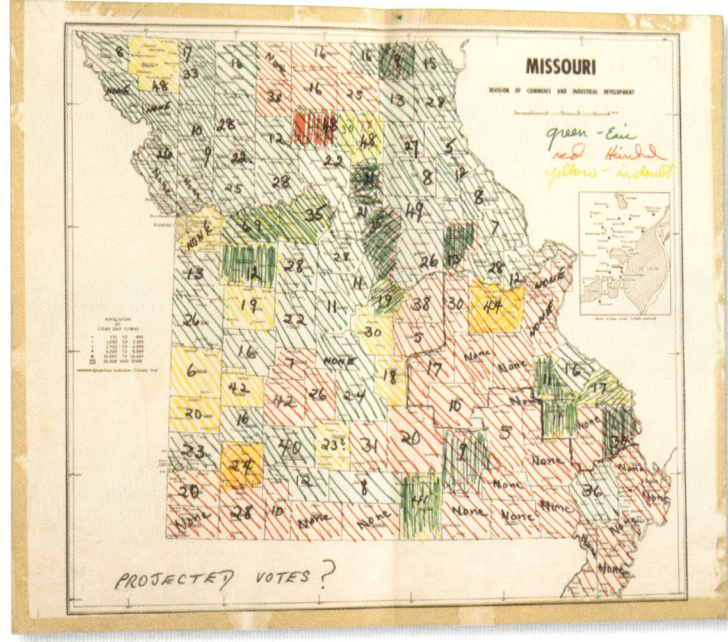

▲ After winning the vote, Thompson addressed the convention, asking for applause for Fred Heinkel's dedication and years of service.

▶ This photograph was taken immediately after Eric Thompson ascended the podium. The photographers surrounding him are from news services. A stunned silence filled the room with news that Heinkel had been defeated.

◀ **Opposite Page Top:** Fred Heinkel counted down the minutes until the election in the Hearnes Center at the University of Missouri. Ray Young is on the right. Seated beside Heinkel in the audience, Young would deny telling Thompson he would stay on as vice president. It was an uncomfortable moment, Young would recall. Young was re-elected vice president.

◀ **Opposite Page Bottom:** Eric Thompson was a master of both strategy and tactics. In his 1979 campaign for MFA's presidency, he carefully mapped out counties and delegates needed per county.

Toxic board atmosphere

Eric Thompson's troubles had just started. He had been elected president of an association/organization/cooperative that, despite profitability, had a fiscally shaky infrastructure and a reporting structure bankers described as baffling. Overseeing that structure was a board of 36 farmers who eyed the newcomer with emotions ranging from outright contempt to simmering suspicion. Even with seven new supportive board members, Thompson faced 29 other board members. A few flat out hated him for defeating Heinkel and said so. Several more considered his actions traitorous.

"You can't believe the hostility and hatred in the board room," Thompson would say years later. "People walked in and steam was just coming out of their ears. They hated me. Absolutely hated me."

In executive session at that first meeting, Thompson optimistically proposed 21 items ranging from a management philosophy of participative management to offering a meeting process with members by district to explain MFA's actions going forward. For that to be productive, Thompson proposed both he and a board member representing the district meet with MFA members, gather questions and ideas, and report back to the full board for consideration. Thompson also told the executive committee members that he saw need for a full-time vice president and full-time executive vice president. Young was performing both of those roles in addition to his job of running MFA Oil Company. That was too much to expect of one man, said Thompson.

Thompson told the members he wanted Young to continue representing MFA on the boards of interregionals and also have responsibilities for day-to-day financial activities. Another proposal was to form a management council of three (himself as president, Young as vice president and an executive vice president to be chosen by a search committee). Here's direct testimony to Thompson's fearlessness. He wanted that executive vice president to be no other than Bud Frew and said so.

"When I told the board that I was going to strongly go after bringing Bud back in," said Thompson, "they tried every way in the world to stop me. They had a board majority against bringing him back. But I brought him back because I knew we had to have him. I couldn't do what I did without him." Thompson freely admitted he needed Frew's financial expertise. "I didn't know the financials like Bud did," Thompson said. "But Bud knew he needed me when we went out to meetings and farmers would be all over us. I could handle that. I knew how to deal with them."

With no time to celebrate victory, Thompson and Young took to the membership to explain MFA's financial difficulties and how they planned to overcome adversities the worsening economy was supplying almost daily.

Thompson was successful, bringing Frew back by the Aug. 28, 1979, board meeting as vice president of retail distribution. It wasn't until the May 20, 1980, board meeting that Young would recommend Frew to be vice president of operations.

Next up at that first board meeting after the convention were two representatives from the St. Louis Bank for Cooperatives: Jack Harris, vice president, and Doug Sims, senior vice president. They were straightforward in their remarks. Stability at the cooperative was important, Sims said. The elections "will make MFA a stronger cooperative." But this meeting, right now, said Sims, "is a most critical date in the history of MFA. In setting a new direction for MFA, the board must be involved," he continued. "Differences must be settled in the board room and not outside in the newspapers, coffee shops or on the highways of Missouri." The bankers would later have more extensive, private talks with Thompson, Young and Frew.

Asked by a board member to comment on the day's election and Thompson's ideas, Young said he had been executive vice president for several years in name but was not totally free to operate as such. He pronounced Thompson's ideas to be sound. Young told those assembled he hoped Bud Frew would be selected and brought in quickly. He could work with Bud Frew. But Young wanted an outside firm to study MFA's structure and make recommendations before any major changes were undertaken. Sims seconded that motion. His bank, in fact, had previously recommended the study while Heinkel had been president. But more importantly, Sims said, "The Board has to go on record for a commitment to change." And the board so moved. In fact, at the urging of both Young and Sims, the board voted unanimously to publicly support the new president. The key word would be "publicly."

At the next board meeting on Aug. 28, the board rehired Frew who would, in turn, bring in David Jobe as vice president of finance that fall. MFA, to Thompson's and Young's chagrin, had never had a chief financial officer who was free to make decisions and provide in-depth analysis. Bill Streeter, today MFA president and CEO but at the time heading up the farm supply division, was immediately impressed. "When Jobe came in to manage finance," he said, "we had no idea of cash management, targeted gross margin, inventory turns, balance sheet metrics. Everybody had looked at numbers, but never at this level. It was new to the company."

"The Board has to go on record for a commitment to change."
—**Doug Sims**

MFA Stores

Pleasant Hope

Republic

Rhineland

Richland

Rogersville

The board also authorized Touche Ross and Company to conduct an organizational study. At the meeting of the board on Nov. 5, Touche Ross officials issued a wake-up call. Although officially looking at corporate structure, the firm was immediately "concerned with profitability of MFA, that the balance sheet was not as strong as it should be, and the cash flow and working capital are not what they should be."

Touche executives found overlapping functions, poorly defined jobs, no company plan for the future and no backups for employees about to retire. They had three immediate goals: 1) improve the financial condition and performance of the company; 2) run MFA like a business; and 3) improve the economic position of farmers.

At the next board meeting on Nov. 26, Touche executives were back. Their findings? More trouble. The association did not have a proper financial position from a profitability standpoint and the balance sheet was inadequate; there had been inadequate emphasis in recent years on profitability; there had not been a full-time manager for day-to-day operations; politics had played a major role at the top of the organization in management; MFA had made inadequate use of modern technology; a corporate long-range plan had not been developed; and potential turnover in key management people could become a problem.

More ominously still, interest rates stood at 13 percent on operating money. Rates would only hover there briefly before starting a precipitous and calamitous climb to 20 percent.

Thompson and the new board members began bylaw changes on term limits for directors, expanded the number of board meetings from four to six, mandated no person be elected to the board after age 70, mandated the president and vice president be elected for three-year terms not to exceed four consecutive terms unless authorized by 60 percent of the delegates, made the president the chief executive officer of the association, added the executive vice president position with duties of the CEO but subject to the president and board, and raised earned membership from $250 to $1,000. They were starting to accomplish things.

Then a landmine exploded beneath them.

That landmine was the MFA Insurance Companies. Reigniting the embers of tension in the MFA Incorporated board room, MFA Insurance Companies had quietly changed their policy from the MFA Incorporated board having voting proxies to those proxies being given to insurance company officers. With the death of A.D. Sappington, in the spring of 1979, Howard Lang became president. The process and precautions were rational but would have been preventable had MFA in years past taken a more fiscally sound approach to structuring companies. The lax structural policies of Hirth and Heinkel came home to roost.

MFA Insurance Companies stood as a separate business. Insurance executives wanted to keep it that way. They worried about changes at MFA, and more importantly, about the aftershocks of Thompson's election. They wanted no part of the upheaval. They had a huge customer base and assets to protect. Years later, Lang would explain it simply: "We didn't want that s.o.b. Thompson getting his hands on our money."

Several Thompson supporters wore this t-shirt in subsequent elections to show enthusiasm for their candidate.

A few current corporate board members of MFA Incorporated still sat on the board of MFA Insurance Companies and had prior knowledge of the actions. Despite his recent defeat, Fred Heinkel was still president of the board of directors of MFA Insurance Companies. MFA Incorporated's attorney advised immediate action to keep from losing control of one of MFA's most profitable enterprises. The dual board members and other seasoned board members counseled caution. Thompson and company then made a serious mistake at a time when even Thompson said there was no room for error. MFA submitted both the names of Thompson and Young for election to the insurance companies' board, which immediately sent up red flags to insurance executives, as well as to a wounded Heinkel. Sensing potential takeover, they bolted, eventually giving up the MFA shield and name. MFA Insurance Companies would become Shelter Insurance Companies.

Back in MFA's board room, the insurance situation slowed progress and sped discord at the next few board meetings, so much so that Young, uncharacteristically, found himself lecturing the board. Young saw in the meetings "an intolerable working situation" that needed to be fixed and fixed now. Young knew important decisions had to be made and made quickly and dispassionately. He quoted the writer James Thurber at a board meeting: "Do not look back in anger nor forward with fear."

Illustrating two problems facing the board in late 1979 and early 1980, this plea from the MFA Incorporated board to Fred Heinkel was signed by all officers with the notable exception of Edwin Sachs. Sachs was also serving on the MFA Insurance board. Sachs would challenge Thompson for the presidency in 1980. (Western Historical Manuscript Collection)

Proud Past, Bright Future: MFA Incorporated's First 100 Years 149

This is MFA Insurance Companies' home office building in Columbia in the early 1970s. By the early 1980s, the company had adopted its new name of Shelter Insurance Companies.

Young tried valiantly to get both new and old board members to focus on the gravity of the situation in-house and the need for board cohesion. "Those high earnings from fertilizer may continue through the calendar year 1980, but with farm prices as low as they are, there is a distinct possibility that the use of fertilizer will go down as farm income goes down and if money dries up to finance production costs," he said in a spring board meeting. The insurance company situation is a distraction, he said. Focus on fiscal survival. He pointed out,

> On top of this, as we well know, high interest rates are killing us. The increase in interest costs compared with one year ago on borrowings from the St. Louis Bank for Cooperatives is now running at an increased cost (not total cost) of $750,000 per month, or nearly $10 million per year… Other inflationary costs, including labor costs, are going to increase this coming year by as much as $3 million. Summing it all up, if our patronage refunds from interregional fertilizer cooperatives come down to an average of what they have been over the last 10 years (and I don't know of any other way of predicting what they are going to be in the future), it could be very feasible that we could be operating without earnings in MFA Incorporated.

Young would then cut to the chase. And he got the entire board's attention: "It is difficult not to look forward with fear when you think of what the consequences of that would be. It would effectively mean no more bank borrowings—no expansion of any kind. We would be a target for a take over."

"It is difficult not to look forward with fear when you think of what the consequences of that would be."

—Ray Young

150 | TRANSFER OF POWER

Unfortunately, at just this time a frustrated Fred Heinkel was aggravating the situation behind the scenes. Documents in his papers at the Western Historical Manuscript Collection show considerable orchestration with several members of MFA's board. The documents are not in his handwriting, so the extent of his involvement is unknown, but his possession of the papers with notes and strategies (and detailed copies of minutes from each MFA Incorporated board meeting) speaks volumes. He could have stopped the subterfuge in its tracks with one stern word. Behind-the-scenes manipulation took the form of having one disgruntled member introduce motions to stop Thompson initiatives and another to second the motions, down to the where's, when's and how's. The papers cover more than a year's period of strategies to thwart Thompson.

Unaware, Young would go on to tell board members that support of management was essential. That board members needed to understand their role was not manipulation of the cooperative's management but oversight in direction. "We can't afford to have friction either within the board of directors or between board and management," he said. "I doubt if you as the board want to take over making the daily management decisions in running MFA Incorporated. Perhaps this is our fault, but we do need a better and more clear-cut definition between policy-making, which is the responsibility of the board of directors, and management of the day-to-day operations of the company, which is a function of management… Every single man on the management team feels that they cannot produce in the climate under which we are now working."

Eventually the board would coalesce. Stouffer today argues for understanding the environment of the times. These men, Stouffer said, worked with Fred Heinkel, knew him and respected him. Most of those men, said Stouffer, were doing what they thought was right. Once they realized we weren't there to wreck the organization but to save it, they worked with us, said Stouffer, who even today lists several of those initial board members as good friends.

Keep in mind, he said, these men only had had four board meetings a year. "They were in there and out in under two hours," he said. They were operating with inadequate information. "The renegades knew more about the business than the board members did. The board members had no idea what was happening or how they were being steered and manipulated." Stouffer and Thompson made it their mission to extend the board meetings by hours (and even days) and to fully inform everyone of fiscal issues. Within a couple of years, Stouffer would be elected board chairman.

"The first year was rocky," said Stouffer, "because we couldn't win a vote. But after that when we got new board members in, like O.D. Cope from southwest Missouri, most of the existing ones realized we were there for the good of the company, that our hearts were in the right place." The board would make many important decisions in 1980. Appointing Don Copenhaver as vice president of retail operations was one.

SOMETHING OLD . . .

M·F·A INSURANCE

SAME THING NEW

SHIELD OF SHELTER

Proud Past, Bright Future: MFA Incorporated's First 100 Years   151

At a district meeting, Bud Frew laid out challenges facing the cooperative. Frew and Thompson split speaking duties, but at all times, both men tried to keep the membership informed by announcing decisions and strategies.

★★ Chapter 8 ★★
The Agricultural Depression

A turbulent, but necessary reign

Thompson would be opposed at the 1980 MFA convention by corporate board member Edwin Sachs and a third candidate. Thompson, however, would win handily. MFA members elected Thompson with 80 percent of the vote and Young with 85 percent, a needed boost of confidence and a loud statement to current board members that the membership was behind Thompson and Young. The intense board fights were over.

Within four months, U.S. President Jimmy Carter decided on "the avenues of peace" to enact a U.S. grain embargo to punish the Soviet Union. Exports ground to a halt. Agriculture reeled. So, too, did MFA. Carter had promised no embargoes. In fact, in 1978 while speaking at the MFA convention, Carter made that very pledge. Unfortunately for U.S. farmers, Argentina was more than happy to access the newly opened Soviet market. The prime rate hit 20 percent. Inflation began its steady march past 15 percent. Farm income fell more than 30 percent.

MFA officials launched the company's strategic planning in earnest. After the 1980 MFA election, Bill Stouffer was elected to the board's executive committee and pushed management hard on strategy.

The U.S. economy continued its downward spiral. Farmers Export Company, ownership of which MFA held with several other cooperatives, lost, at first report, $23 million in continuing grain crises and mismanagement. MFA's share of the loss approached $1 million. Far-Mar-Co, another grain giant, attempted to purchase the struggling grain co-op.

MFA pushed hard to restructure by consolidating locations and shuttering others. "Our beginning criteria was," explained Thompson, "had that exchange lost money every year over a five-year period? And we still owned them? That's the way I approached farmers. Are you going to keep owning something that's losing money every year and keeps getting worse? No, you're not. So don't ask us to."

"Are you going to keep owning something that's losing money every year and keeps getting worse?"

—Eric Thompson

▲ Bud Frew maps out his main points prior to the convention at the Hearnes Center in the early 1980s. While focused on improving operations and meeting the challenges head on, Frew had identified locations to be consolidated or closed.

▼ Thompson and Frew present facts to the membership. The two men crisscrossed the territory explaining how their decisions were necessary for MFA to remain a viable cooperative for future generations.

Thompson took the first rounds of meetings with local board members when MFA was closing a facility. "They knew, down deep they had to know what was coming," he said. But it was emotional, he said, like he was closing a school, church or a community. "In some cases," he said, "the community actually dissolved. I hate to use a Clinton quote, but I did feel their pain." Frew would describe the existence of locations in terms of when they were built. Most had been positioned when teams of horses or mules pulled wagons of grain. That explained proximity.

MFA consolidated Conception Junction and Ravenwood; Downing and Lancaster; Trenton, Laredo and Spickard; Wyanconda and Williamstown; Blockton, Sheridan and Grant City; the list went on. Simultaneously, MFA closed Beaufort, Bunceton, Greenfield, Linneus, Purdy, New Haven, Mountain View, Eugene, East Prairie, Warsaw; and that list goes on as well. Bulk plants were reassessed and closed if they were redundant. MFA sold some exchanges to area

154   The Agricultural Depression

From left: Don Copenhaver, Eric Thompson, Keith Mitchem and Bud Frew discuss strategy during the early 1980s. With a prime rate approaching 20 percent, a Carter-imposed grain embargo and a steady upward march in inflation, MFA executives faced daunting challenges.

farmers. Sometimes those farmers made successes out of businesses MFA executives had written off—Meta stands today as a shining (and profitable) example.

"We had such a concentration of locations," said Stouffer. "It was common to see locations eight miles apart. Our biggest competition was internal competition between MFA locations, whether locals or company owned. It made no sense. Look at Marshall and Slater, how close they were. I know we had built fertilizer plants for political reasons. We had one in Sweet Springs, we had one in Emma, we had one in Concordia. They were just a few miles apart. Again, when you put rubber wheels under agricultural loads, it's a completely different situation."

As Bud Frew would write in a 1981 memo to Copenhaver: "We need to sit down at our earliest convenience and define as closely as possible both the company-owned as well as local exchanges that will be closed or consolidated within the next five years. This should be a rough cut for the purpose of explaining the program to the bank."

Frew was trying to improve operations, not necessarily impress the bankers. Still, he knew lenders ultimately controlled the cooperative in view of current debt levels. MFA's operations and basic structure had to improve. Frew wasn't anywhere near finished. He and Thompson were just getting started. The bankers had scheduled a meeting with top management and members of the board for Sept. 30, 1981. Frew would be prepared.

MFA Stores

Rolla

Rosebud

Rumpus Gap Gang

Russellville

Salisbury

The St. Louis Bank for Cooperatives had been blunt with Thompson and Frew. MFA had never operated with a strategic plan. It was a glaring weakness for a company of MFA's size. Thompson and Frew were committed to making a strategic plan an integral part of MFA's operations. MFA has operated with these types of plans in place ever since.

Time to face facts

Harry Chelbowski and Doug Sims had hard facts to present. Their challenge was to present those facts without alienating their audience. It was essential those facts and implications be absorbed. Chelbowski was president of the St. Louis Bank for Cooperatives; Sims, senior vice president. Thompson, Frew, Young, Jobe, Copenhaver and several other MFA executives as well as a handful of board members met to listen to lender concerns.

Sims cut to the chase. "We are going to have some pretty frank comments and would ask that you not over-react to some of that frankness but question anything you might be concerned about or you don't agree with." Chelbowski and Sims proceeded to lay out the financial landscape of midwest agriculture in the 1980s and MFA's position within it. To start, said Chelbowski, everyone in the room needed to completely understand that as it now stood, MFA was too highly leveraged. There was far too much debt relative to equity.

Historically, MFA had suffered from low to no profitability—at least in terms a businessman should be comfortable with, Chelbowski said. "Over a large period of time, this company has basically not been

---

PREFACE

Strategic planning for any corporation conducting business in today's environment is essential. Without a carefully thought out and articulated plan that identifies the precise direction the organization must follow, the resources available, and the priority of useage for those resources the organization will in all likelihood fail to remain a viable entity.

The process begins by corporate development of a strategy or strategies that will lead the organization, govern the expenditures of resources, and measure the results of the progress made toward attaining the objectives set.

Until corporate strategies have been developed, it is impossible to ask those of you with more specific responsibilities to develop strategies and the guiding plans for your own key areas.

We now feel that those strategies have been established corporately and will provide the guidelines that will allow each of you to make the kind of contribution necessary to complete the process.

The process can be difficult and time consuming, but it will always be very rewarding if quality thought is put into the endeavor.

The time that each of you spend on this process will be the most valuable time you will spend for the corporation.

B. L. Frew
Executive Vice President
MFA Incorporated

profitable on average," he said. Why? Low liquidity—not enough working capital. MFA had struggled just to pay off seasonal loans. MFA had tried for too long to be everything to everyone.

Chelbowski could not believe an organization the size and complexity of MFA elected a CEO from the membership each year. In all his years, he had never seen any other organization that would subject itself to such risk. He'd made that speech before to Fred Heinkel.

Sims interjected. If a candidate were to run on lower input prices and higher grain prices, unless the membership is very sophisticated, he'll be elected to liquidate the cooperative. The easiest and yet most destructive thing you can do, he told the assembled executives and board members, is to pay the farmer higher prices for crops and charge them less for products. If you think that is MFA's role, you will liquidate MFA a ton, a bag and a bushel at a time. Professional management from here on out is an absolutely necessity.

Both men counseled continuing to listen to the farmer. But management and the farmer needed to understand one thing from the start. MFA could not provide a service at a loss, no matter how popular, and the farmer should not expect MFA to. Not just co-op management, but their customers needed to understand that basic business principle. From their perspective as primary lenders, the bankers listed MFA's historic weaknesses:

1) Financial Structure: Too much debt; too many non-productive assets; no real financial plan
2) Operations: Inadequate profitability and cash flow; lack of marketing and pricing philosophy; totally service oriented
3) Organization Structure: Too many MFA organizations for adequate control; organization chart doesn't fit the business; board president (CEO) elected by membership; weak board of directors
4) Management: Too many political decisions; lack bottom line orientation; weak financial planning; inadequate controls—too many surprises
5) Competition: Strong cooperative and private competitors; lower cost structure
6) Local Cooperatives: No strong leadership from MFA; too many weak locals
7) External Relationships: MFA in turmoil; lack good relations with other cooperatives and farm organizations

Now is not the time to assign blame, the bankers warned. Now is the time to commit to moving forward. The lenders stepped back and waited for questions. One of the first questions came from a board member who had served under Heinkel. It was the first time he had seen or heard of these problems.

Why hadn't they been told this years before? Sims's reply? "We did." Both Sims and Chelbowski had made the above recommendations to Fred Heinkel repeatedly, the last time in 1978. "It was not well received," said Sims. "Almost nothing was done with it for a year. It put us in a very, very awkward spot." With the contested election approaching, the bankers held off and waited for the outcome. Raising the issue during the election year would have been toxic for all involved. So they waited.

If Heinkel had been re-elected, the bankers were determined to take the situation to the full board whether Fred Heinkel objected or not. As it was, the bankers had met with Thompson, Young and Frew just a few months after the 1979 election. They outlined the same concerns and were gratified to receive immediate attention and, more importantly, complete agreement. It was a start.

MFA was not in this alone, the lenders said. They had studied 18 farm supply cooperatives in the last five years. They had seen deterioration in loan quality across the board. They found increased leverage, reduced solvency and reduced liquidity. Cooperatives had not responded well to the new operating environment. It was not a cause for panic, Chelbowski and Sims said. It should be a call to action. If cooperatives would take the necessary steps to improve their balance sheets and operations, they would survive the current bleak financial landscape and re-emerge after five or so years as healthy, profitable businesses. But only if corrective action was taken. Those unable to adapt would not survive.

Too many cooperatives had been expanding, said the lenders, using exorbitant interest rates in risky areas like grain. From 1972 to 1981 MFA's working capital went from $16.7 million to $25.3 million. Meanwhile sales increased from $214 million to $872 million. So working capital increased approximately 1.5 times while sales increased four times. Those who make their living on balance sheets, the lenders pointed out, expect working capital and sales to march in tandem.

Sims and Chelbowski also recommended MFA officers look closely at the unwieldy structure of the cooperative. MFA had almost a dozen interlocking affiliates, held together by a political structure of Class A and Class B board members. The litigious atmosphere of the 1980s made those associations dangerous. Lawyers always looked to as many sources of revenues as they could find. That made MFA a big target for litigation. MFA Livestock Association and Farmers Livestock Marketing Association were two such concerns. Thompson, Frew and Stouffer analyzed those relationships closely and recommended beginning steps to limit liability.

By 1980 MFA's long-term debt was $90.2 million. MFA had relied too heavily on debt to finance growth, and much of that debt ($66 million) was bonds. Most of those bonds had a five-year redemption privilege. If every bondholder whose bond dates fell within the five-year timeframe were to walk into MFA and demand money, MFA would be liable for $13 million cash in the midst of a liquidity crisis. MFA's managers and board members literally cursed out loud. Sadly, a variation of this same scenario would bankrupt Farmland Industries in the future.

One of the original board members lightened the somber mood during the bankers' recitation of a weak board. He would admit, he said, to having been a weak board member. But he made up for it by being loud. Fortunately, the days of passive boards had ended at MFA.

"Somebody's got to pay the piper"

Far from dispirited, Thompson, Frew and the executive committee were re-energized. With complete board and lender support, they were on the right track. Affirmation from the officials at the St. Louis Bank for Cooperatives was essential for progress. What's more, they had a plan that corresponded perfectly to the concerns pointed out by the bankers. Still, the hard realities of economics tempered their moods.

From 1971 to 1981 average U.S. farmland prices had soared 266 percent. Land values could easily, and disastrously, reverse course. Economists were, in fact, predicting it. Farm debt had nearly doubled in five years. Interest rates had continued their climb. As Frew would say years later, an investment that looked great at 6 percent, looked like hell at 20.

Thompson and Frew cut MFA's workforce by $1.4 million in payroll, started in earnest to reduce "non-income-yielding" assets that did not return acceptable levels of profitability. They continued to close or consolidate locations with "overlapping or inadequate" market territories. Thompson would tell the membership, "Ours is not a philosophy of maximizing profits at the expense of MFA members. It is, rather, the application of acceptable profitability measures to assets employed and sales generated that are necessary to meet your projected demand for products and services." And still, MFA lost money.

The pain suffused ag industry. Farmland Industries lost $55.6 million in 1982 and estimated losses as high as $62 million for 1983. Farmland trimmed its workforce, planned sales of divisions and halted joint ventures. The U.S. economy soured. Agriculture was moribund. For 1982, MFA, despite a common plan, aggressive cost-containment measures and board consensus, would post a loss of $15.7 million, following 1981's profitability of $8.5 million. MFA, officially, was in violation of existing loan agreements with the St. Louis Bank for Cooperatives.

Media jackals shrieked, did their level best to drag down MFA executives, rooted through dumpsters for documents and extensively quoted unidentified sources demanding Thompson's head. Thompson would announce: "1982 has been a most difficult year for MFA members and their cooperatives. But by adhering to tough, prudent and sound management actions, MFA and its members continue to exemplify the resilient heritage of the American farmer." As Thompson would go on to explain, the worst was not yet over.

Frew brushed off criticism and moved forward. "Our farmer-member expects MFA to perform just as well as the private sector," he said. "And, ultimately, he expects us to do it in a way that will enhance his profitability. We think we need to move toward a marketing and sales-driven organization that does not create a need, but meets a need." Frew was all in.

MFA Milling Company executives asked MFA Incorporated to take over its operations. The Springfield-based feed business had prospered for 50 years. But the livestock industry had changed dramatically and for good. Milling Company assets were almost exhausted. Updating equipment and facilities would cost hundreds of thousands if not millions of dollars. The MFA Milling Company coffers were close to empty.

As strong as it had been, its operations highlighted a basic contradiction in MFA's overall structure. MFA Milling Company stood on its own with a rigidly defined trade territory. Simultaneously, MFA Incorporated had its own feed operations. Back in Hirth's days, St. Joe Feed and Grain had supplied north Missouri and MFA Milling had supplied south. But four decades had eroded those boundaries. MFA, again, found its worst competition was internal. MFA Milling and MFA Incorporated feed salesmen competed for the same customers, resulting in lower prices and inadequate margins. The relationship stood as a shining example of what lenders called obstacles to adequate control. MFA Milling Company was run for almost 40 years by John F. Johnson, an able executive. But Johnson and Heinkel shared the drawback of hubris. Neither would give in to the other.

As the St. Louis bankers would point out, Johnson's objective was to have the lowest cost feed. "Over time," said the bankers, "it sold its business a bag at a time." When MFA Incorporated finally absorbed the company, much of its operation would be shuttered, other portions would be enhanced and repositioned. To make those business analyses, Thompson, Frew and Stouffer sent David Jobe. As Stouffer pointed out, they could count on Jobe's judgment.

His analysis revealed the operation had exhausted its assets while not retaining adequate earnings. Almost all profit had been returned through patronage to customers (82 percent cash patronage). A cooperative, said Jobe, cannot operate that way from a business or ethical standpoint. That shortsighted practice benefits today's customers at the expense of their children. When the assets are gone, the business is gone. MFA would take over remaining assets, but at heavy cost.

Losses continued into 1983, with MFA ending the year with a negative of $3.5 million, almost a third less than 1982. With the economic horizon dark and MFA's financials still declining, the St. Louis Bank for Cooperatives began urging MFA to consider merging with Farmland or Growmark.

Mergers from a position of weakness are recipes for disaster. Thompson and Frew

"But by adhering to tough, prudent and sound management actions, MFA and its members continue to exemplify the resilient heritage of the American farmer."

—Eric Thompson

balked. Frew identified more exchanges to close and seven additional bulk fertilizer plants with returns on assets a negative 7.9 percent. He looked closely at closing feed mills and farm supply warehouses. He listed interregional investments he wanted to shed. He and Thompson reorganized the Columbia home office. Frew pushed to sell MFA's soybean processing plant in Mexico. "The best it has ever done in its best profit years was 11 percent return and the average was 2.6 percent return, and if that is not an asset that is not yielding, then I have never seen one," Frew said.

Jobe, who would handle much of the analysis in selling the facility, agreed with the decision forced by fiscal realities of the early to mid-1980s. Looking back, though, he regrets the move. The business had tremendous potential, he said. What it had needed was good management. MFA, at the time, had too much on its plate for the management team to spend several years focusing on the plant's profit potential in the midst of a fiscal crisis, he added. It was a decision economics forced.

Stouffer, as chairman of the board, hastened to tell MFA members there was absolutely no question about the security or safety of member grain. By November of 1984, Stouffer explained MFA had received an unqualified audit saying the auditors were not unduly concerned about MFA's viability. Stouffer said,

> Equity as a percent of total assets is less than desirable—but then, it always has been. In 1970, members' equity as a percentage of assets was 14.8 percent. That equity ratio peaked at 27.3 percent in 1976. It decreased to 19.6 percent in 1978, and today, our equity remains at 19.6 percent of assets. Double-digit interest rates since the early 1980s, coupled with our traditionally low members' equity ratio and our unproductive interregional investments have prevented us from making the kind of improvement we would like to have seen in profitability. Once fully implemented, however, our consolidation plan will help us improve our profitability and member equity position in MFA.

Bankers Chelbowski and Sims still saw merger with another cooperative as a strategy to strengthen both partners. MFA was not the bank's only customer suffering. Growmark had no desire to acquire MFA's debt. That left Farmland. So Stouffer traveled to Arizona to meet with board members of Farmland Industries. It chapped him. "They just had that attitude," he said. "They were too big to fail." Farmland proposed merging with MFA and keeping MFA as a division, complete with current logo—with updates.

The two cooperatives were a bad fit from anyone's perspective. MFA had competed head to head with Farmland all of Stouffer's life—and before, clear back to Howard Cowden days.

"Most of that competition had to do with egos," he said. "It was an inane, insensible competition. MFA would put up a location and Farmland would put one right beside it. It made no sense. Look at the locations. Look what happened. The membership hadn't figured out it was their money they were playing with and they allowed it to happen." The only difference Stouffer could see was that Farmland tended to be Republican and MFA Democrat.

Stouffer came back from Arizona and told Frew, "There's no way we want to get involved in that. They are where we were. We're not gaining anything. They need to make some of the same changes we've had to make but they think they're big enough they don't have to. It's just the same thing all over again."

Years later, reflecting, Stouffer would observe, "Farmland never made those decisions. I don't care how big you are. Sooner or later somebody's got to pay the piper. We'd already made the tough decisions. But they wanted to be an international exporter and had no idea what they were doing. Just like the federal government today."

Still, MFA ended 1984 with a loss of almost $5 million. The St. Louis bankers pushed merger talks with Farmland front and center. Thompson made his case for merger in an August/September 1984 editorial in Today's Farmer. By the October issue he was in full backpedal. The deal was off.

Eric Thompson never ducked an issue. He had campaigned for MFA president on a platform of limiting board director terms and promised to serve only six years himself, if elected. "I wasn't going to go back on my word," he said years later. "Not only did I make [the promise] to myself, I made it to thousands of MFA members."

MFA Stores

Shelbina

St. James

Ste Genevieve

Stover

Sturgeon

Promises to keep

Running for the presidency of MFA in 1979, Thompson had committed to serving only a six-year term. Thompson's term expired in 1985. Technically, Thompson could continue if the membership voted by more than 60 percent to retain him. Another matter stood in his way. Honor. "I wasn't going to go back on my word," said Thompson. "Not only did I make it to myself, I made it to thousands of MFA members. Not one in a hundred CEOs would have had the [courage] to pick up and leave like that." Thompson did.

Thompson visited extensively with his management team and the board. One of his key proposals had been to have the board of directors hire and fire the president rather than having the president subject to political forces, whether every year or every three years. That proposal became an MFA bylaw at the 1983 convention.

Simultaneously, Thompson argued, if the president was the chief executive officer of the cooperative who was responsible, along with the board, for MFA's direction, why also employ an executive vice president who essentially had the same duties? The timing was perfect from Thompson's perspective for combining the positions of president and executive vice president. That meant either Thompson or Frew would be out of a job.

Board Chairman Bill Stouffer would announce the elimination of Thompson's position effective Jan. 1, 1985.

MFA Incorporated

201 South Seventh Street
Columbia, Missouri 65201
314-874-5111

October 30, 1984

TO: All Employees

At the meeting of our MFA Board of Directors today, I announced my resignation as president and chief executive officer as part of the consolidation plan that was shared with you recently.

I would have preferred to make my decision known to you at that time, but felt I was ethically bound to take my decision first to the board.

The plan that I presented to the board included the recommendation that the positions of president and executive vice president be combined; therefore, considering the role and responsibility of the CEO of the consolidated company, I have decided to step aside. My resignation will become effective January 1, 1985.

When I was first elected president in 1979, I promised MFA members that I would not stay in this position for more than six years. I also shared with them my goals for the company: first, to change MFA from a quasi-farm organization to an agribusiness cooperative with emphasis on operations; second, to depoliticize not only the selection of MFA president and board of directors, but the entire company; third, to restructure MFA operations in a manner that reflected the realities of the changing agricultural economy. I believe that I have accomplished each of these goals.

The last four years have been both agonizing and difficult for management and employees of MFA. I know personally the anguish associated with the actions that we have had to take, yet I've always believed that I must apply the same standards to myself that we apply to all employees.

I respect this company and all its employees. You are a group that is second to none. Despite any personal impact, I believe I must take this action to insure that the plan is implemented in detail and will result in the strongest corporate structure possible after consolidation.

One of you asked me the other day if I thought MFA could survive in light of the continuing declines in the agricultural

"I believe that the successful implementation of the plan approved by the board will assure the basis for profitable performance on the part of MFA — even under adverse weather, economic, and federal farm policy conditions."

—Eric Thompson

-2-

economy. I believe it can. I believe that the successful implementation of the plan approved by the board will assure the basis for profitable performance on the part of MFA--even under adverse weather, economic, and federal farm policy conditions.

    I feel confident that I have served you and this cooperative well and with all the energy I possess. I have appreciated your continued confidence in me during these difficult times.

    I am pleased to inform you that the MFA Board of Directors has selected Bud Frew to assume the responsibilities of the Chief Executive Officer effective January 1, 1985. Please join me in wishing Bud the very best.

                                Sincerely,

                                Eric G. Thompson

mk

Not only did Thompson follow through on the promises he made during his campaign, he began streamlining the organization to the extent of eliminating his own position. MFA, he noted, did not need both a president and an executive vice president. He combined the two roles and left the company.

"When I was first elected president in 1979," Thompson wrote in a memo to all MFA employees, "I promised MFA members that I would not stay in this position for more than six years. I also shared with them my goals for the company: first, to change MFA from a quasi-farm organization to an agribusiness cooperative with emphasis on operations; second, to depoliticize not only the selection of MFA president and board of directors, but the entire company; third, to restructure MFA operations in a manner that reflected the realities of the changing agricultural economy. I believe that I have accomplished each of these goals."

In that same memo, Thompson addressed the issue on everyone's mind: the survival of MFA in the face of the continuing agricultural crisis that future economists would later label an agricultural depression. "One of you asked me the other day if I thought MFA could survive in light of the continuing declines in the agricultural economy. I believe it can. I believe that the successful implementation of the plan approved by the board will assure the basis for profitable performance on the part of MFA—even under adverse weather, economic, and federal farm policy conditions."

He left out the two main reasons MFA would survive. Bud Frew and Bill Stouffer.

B.L. Frew presided over 12 straight years of profitability at MFA Incorporated, an unprecedented feat. Frew made sure MFA's financial house was in order. After instituting financial stability, Frew began building a legacy of competent management and honest communication.

★★ Chapter 9 ★★

Modern Management

"We'll do it our way"

No shrinking violet, Frew would tell MFA members to forget the rumors in the countryside. But keep this in mind, he said. MFA still has the largest fertilizer market share in the state. MFA still produces and sells more animal feeds than any other feed company in Missouri. MFA still originates more farmer grain than any other single agribusiness in Missouri. MFA exchanges still comprise the best ag-products distribution system in the state and offer the most comprehensive, one-stop service centers. So why all the hand wringing? It's time to move forward.

MFA had sold its river terminals to Farmland Industries in June of 1984 and its soybean plant the prior year. Neither sale stopped year-end losses. "Be assured," said Frew, "the sale of assets is intended to remove non-productive investments—not merely to 'shore up' the bottom line." MFA had lost $10 million on those grain terminals in the past two years. Talk about non-productive. Farmland, within a year, was regretting that purchase and trying to resell those same structures to yet another cooperative. The grain market of the early to mid-80s was no place for those without tremendously deep pockets.

"It's time to move forward."

—**Bud Frew**

"The simple truth is," Frew wrote in Today's Farmer in 1985, "that farmer-owned agribusinesses are overbuilt for today's highly competitive marketplace." Frew was responding to a spate of mergers in the cooperative world. MFA and Farmland Industries talked about merging but never followed through.

Bud Frew considered Ray Young to be his mentor. The men formed a close friendship over the years. When asked in a 1979 corporate board meeting about the wisdom of bringing Frew back to MFA, Young would say, "I can work with Bud Frew."

Cooperative mergers took place at record levels across the nation. Farm foreclosures led the national news nightly. Congress issued proclamations to save the family farm. "The reason for this seemingly feverish activity is no secret to anyone involved in the board leadership or management of a cooperative agribusiness," Frew would write in Today's Farmer in April of 1985. "The simple truth is that farmer-owned agribusinesses are overbuilt for today's highly competitive marketplace."

Bud Frew and Bill Stouffer dug deeper still into the balance sheet, trimmed all fat and some meat. Spring would show favorable weather and MFA's best spring volume since the 1970s. It wasn't enough. Loss on operations was $2.9 million and another $7.3 million of non-cash loss on discontinued operations. But Frew had staunched the bleeding. Balance sheet metrics inched up. Owner's equity remained at the previous year's 20 percent. Frew announced hopes for moderate profitability by 1986. MFA's market share was still the largest of any agribusiness and was now expanding. That required more cash: $5 million more to be exact. So Frew and Stouffer once again asked the St. Louis Bank for Cooperatives for additional operating money.

The lenders were hesitant. They wanted Stouffer and Frew to meet them at their offices for discussion. They still viewed merger as the most appropriate response to MFA's fiscal woes. Stouffer and Frew waited anxiously and silently in the St. Louis bank offices. Once in the conference room, the two listened patiently while the bank officers explained their concerns. A push to restart stalled merger talks with Farmland could offset balance-sheet weaknesses and guarantee operating loans, said Doug Sims.

While the numbers made the bankers uneasy, the mere thought of them made Frew reach for antacid. He knew to the depths of his being they were numbers MFA had begun to turn around, he would say in the late 1990s. No way he'd allow them to deteriorate after what MFA had already weathered.

"The bankers recommended we write down our bondholders' equity," said Stouffer. "Bud and I were just as determined not to." Bond debt had been a topic of discussion at the bank in previous years. MFA had relied too heavily on that source

Bill Stouffer as chairman of the board and Bud Frew as president and CEO were a formidable team. Both men met with bankers during the turbulent 1980s, listed actions and followed through. After leaving MFA's corporate board, Stouffer would continue to excel, both in his farming operation and in the political arena. First elected in 2004, Stouffer served several terms in the Missouri State Senate.

of cash flow in the past. Both Frew and Stouffer knew those bonds represented farmer faith in MFA. "I can remember when MFA was teetering on the edge and one farmer came into Salisbury and put $30,000 down on bonds," said Stouffer. That's faith. "There was no way we were going to clip the bondholders."

The bankers insisted. Frew's face colored, said Stouffer. MFA's chairman of the board and MFA's chief executive officer exchanged glances. Stouffer nodded. Both men stood. Frew reached in his pocket, grabbed a ball of keys and threw them on the conference table. The keys resounded on the polished wood a little more loudly than either man expected. "We'll do it our way or you take these keys and run the s.o.b. yourself," demanded Frew, before turning and joining Stouffer on the way to the door.

Probably seeing the determination he was looking for, Sims accepted MFA's financial strategy. But more importantly, from Stouffer's standpoint, the keys were returned. As they left the bank, both men still wore their masks of stern determination. Once in the parking lot, both let down their guard. "I'll tell you it was a relief that they bought our deal," Stouffer said. "The keys Bud had thrown on the table were our car keys and only way home."

When Bud Frew retired Jan. 31, 1998, he had presided over 12 straight years of profitability at MFA Incorporated, an unprecedented

MFA Stores

Summersville

Tipton

Union

Van Buren

Always dismissive of media incompetence in covering agriculture impartially, Frew nevertheless became a good MFA spokesman. But he refused to suffer fools. When a reporter complained that MFA hadn't offered a correction for an article another reporter had written, Frew asked why newspapers expected him to do their jobs and further, why they kept incompetent reporters on staff? "I wouldn't," he said.

feat. Small profits of $2.3 million were seen in 1986, but profits nonetheless, followed by $4.4 million in 1987. The upward march would continue. Frew would make the word "profit" acceptable in cooperative circles. "Savings" was a term from the 1920s, he would say.

So too, would Frew focus on changing descriptions of MFA locations. "Exchange" was out. "Agri Services Center" was in. "When was the last time you exchanged eggs for feed or cream for farm supplies?" he would ask any who questioned him. It was the language of a bygone era. Frew focused intensely on upgrading MFA's facilities and service. He also changed the culture of the organization.

"It's our mission, our strategy, our honesty, our integrity. We will do what we say," he proclaimed. "Farmers as a group understand honor. They understand your word. They understand your handshake. We had squandered our reputation in the 1970s. But we've taken it back," he said just before his retirement.

In 1985 total assets had shrunk to $175 million. When Frew retired, 1997 total assets were $295 million. Working capital, which had fallen to $5 million in 1985, had risen to $56 million by 1997. Net worth in 1985 was $32 million and $114 million in 1997. And member ownership, which stood at 18 percent in 1985, had risen to 39 percent.

In retail, Frew focused on providing members with quality products and timely service rather than relying on blind loyalty. "We made a conscious, strategic decision to exploit our strengths and minimize our weakness," he said. "That led us to our focus on service and distribution." That decision would, in turn, lead to those years of profitability.

Frew would remain proud of the path he and Thompson had taken. "Business decisions, first and foremost," said Frew. "We're a farm supply and grain marketing organization. We won't enter into something, we won't expand to please someone else's agenda. We'll pick battles we can win." That accomplishment, he would note, was solely because, "We stopped electing the cooperative's president at large and required the board of directors to hire and fire the chief executive officer." Until that time, he said, the president could look at the board and say, "You were elected by a district. I was elected by the membership. We'll do it my way."

"From that point on," explained Frew, "I could stand up in front of the board and say, 'Fellows, this is right. We're making this decision because it's a good business decision.' We changed the company overnight with that structure."

It remains Frew's legacy.

"Farmers as a group understand honor. They understand your word. They understand your handshake."

—**Bud Frew**

Moving forward strategically

To continue Frew's leadership, the MFA Board of Directors hired Don Copenhaver, senior vice president of retail distribution under Frew. Copenhaver had big shoes to fill and lofty expectations to meet. He would accomplish both goals. An accountant by training, Copenhaver continued and expanded MFA's focus on the balance sheet. But growth was his goal.

Under Copenhaver, MFA would undertake a series of expansions, fueled in part by a desire to counter a weakness in the cooperative business model: access to needed capital. By forming strategic alliances to create private businesses under the banner of MFA Enterprises, MFA helped create AGRIServices of Brunswick, LLC; Central Missouri AGRIService, LLC, in Marshall; West Central AGRIServices; and a series of locations called AGChoice. The choice paid off in spades.

Today, Brian Griffith is senior vice president of operations at MFA, a position overseeing both wholesale and retail operations. For years, Griffith served as MFA vice president/corporate secretary/

▲ In talks with MFA employees prior to his retirement, Frew said, "Remember who owns you. Honesty and integrity are what it's all about. I don't want this organization to ever lose that. Think about what's good for the company, not what's good for my division, for my area."

▼ From Copenhaver's perspective, business growth should come from targeted objectives that meet business goals and specific return criteria. Growth under Copenhaver focused on countering a cooperative's traditional weakness of lack of capital.

general counsel. He's well attuned to MFA's direction, having helped structure and implement it.

Griffith knows accomplishment when he sees it. "Don Copenhaver's support for strategic acquisitions may be his biggest accomplishment. In fact, his willingness and support for ventures, acquisitions and business partnerships have made the additions seamless and profitable. It is recognition on his part that the company needs to grow in a cost-effective manner. I won't call it visionary. That's too grand a term, and Don wouldn't approve of it. But I will call it a common-sense approach to business."

Over the course of Copenhaver's stewardship, MFA's grain bushel sales rose from 42 million in fiscal 1999 to 70 million in fiscal 2008. Seed sales sprouted from $16 million to stand at $47 million. Over the span of Copenhaver's presidency, MFA returned more than $48 million to patrons. In retail volume, MFA Incorporated stood at seventh in the nation in fertilizer sales; eighth in retail sales; seventh in custom application; 10th in seed; seventh in precision ag; fourth in grain sales.

Fiscal responsibility, preached incessantly by first Thompson, then Frew, was a mantra under Copenhaver. Business growth should come from targeted objectives that meet business goals and specific return criteria. "We have an obligation as the management of the cooperative to protect the assets of this company," Copenhaver said.

MFA's culture is shaped to recognize and respond to customer needs. That's where fiscal responsibility comes into play. MFA grew by providing solutions and

"We have an obligation as the management of the cooperative to protect the assets of this company."

—Don Copenhaver

▲ President George W. Bush spoke at the MFA Aurora feed mill in 2002. Bush advisors had called MFA to request the opportunity. In the post-9/11 world, Bush addressed terrorism, trade, agriculture and the farm bill. He also took time to help fill a bag of MFA Cattle Charge.

◄ "If we're talking about the economic health of the country, we've got to always understand it begins with a healthy farm economy," President Bush told the crowd at the MFA Aurora feed mill. Bush called for less stimulus and more private investment and commerce.

Copenhaver made his mark on MFA through intense support for joint ventures, acquisitions and business partnerships that led MFA growth in a cost-effective manner. Bud Frew was publicly proud of Copenhaver's accomplishments and not the least bit hesitant in saying so.

avoiding trendy schemes. As such, its leaders have the long view.

Frew, in retirement observing Copenhaver's performance, was pleased with MFA's accomplishments and understood the pitfalls Copenhaver avoided. "As CEO," said Frew, "it's easy to lose track of your basic focus. Farmers own the cooperative, and they frequently want more than you can responsibly give. It's your job to know when to say, 'No. No, I won't do that. That's not a good financial decision.' The ones you don't sign off on—those are the ones a leader should be recognized for."

Copenhaver would have his share. He had basic operating principles he kept foremost in mind: "Would I do this if it were my own company? Would I do this if it were my own money? I consider myself a fiscal conservative. I do not like undue risk. I want to make sure decisions are made with prudence."

That outlook made Copenhaver miserable the final months of his last year as president. When he'd announced his retirement in summer of 2008, MFA had just posted its most profitable year ever ($45 million). But a U.S. presidential election, weather and an out-of-the-blue worldwide financial crisis pushed MFA to the brink. The problem? Large write-downs in fertilizer inventories were daily peeling strength from MFA's balance sheet—an action that affected Copenhaver as if it were his skin.

Those write-downs were a dose of new market reality to a long-standing business practice. For years, part of MFA's member-service philosophy included having a supply of product. Year in, year out, MFA had historically supplied more than one million tons of dry, liquid, anhydrous and bagged plant foods.

MFA's distribution system, constructed by a generation of cooperative leaders to feed the farmers' appetite, included ocean vessels, river barges, rail and truck from production sites or storage terminals in 12 states and numerous international production sites. Market experts forecast rising prices; MFA filled the system in expectation. And then the bottom fell out of world markets, first in the banking sector and then across the board. The stock market lost 40 percent of its value.

Product didn't go away; it depreciated. Product that wholesaled at $1,000 a ton in August 2008 slid to $500 retail by late December. More than 100,000

Proud Past, Bright Future: MFA Incorporated's First 100 Years | 171

▲ MFA's managers meeting theme in 2009 was "212, the extra degree." The one-degree difference between 211 and 212 is the difference between creating hot water and creating steam to power locomotives. Bill Streeter focused MFA's workforce on creating that steam to power the company.

▶ Opposite Page: Bill Streeter has a habit of having a binder clip (or two) fastened to his shirt front. So employees of MFA's home office hand decorated more than 200 of these custom clips, distributed them throughout the building and sported the clips proudly during his first week as president. Target 20 was Bill Streeter's shorthand for achieving $20 million in profitability.

tons of fertilizer backed up in a system built for rapid turns and began feeding on the balance sheet. And Copenhaver's previously announced retirement arrived in the midst.

The situation devastated Copenhaver. "I didn't want to end my career with a loss year," he told employees in a February 2009 meeting. But he needn't worry. His friend and close associate of many years had been chosen to fill his job: Bill Streeter.

Streeter, in his first column in Today's Farmer, would say, "Don earned his place in history for his balance-sheet management and for his leadership role in focusing the cooperative on fiscal fundamentals." That strengthened balance sheet would be a tremendous asset in what followed.

In the midst of the worldwide market crash, MFA would post a historic loss of $64 million for 2009. But Streeter would prove equal to the situation.

"Planning is everything; the plan is nothing"

When Bill Streeter stepped up to the microphone in the 2009 managers meeting in August, the fiscal crisis and the effects of a loss year were on everyone's mind. Bill Streeter was where he wanted to be: front and center addressing problems head on. "I was hired to be MFA's president and CEO effective March 1, 2009," he began. "I've worked at MFA since 1973. I have faith in you," he paused dramatically with each listing, "I have faith in our customers. I have faith in myself. And I have faith in MFA."

Relief began to bubble up from the crowd as tension began to thaw. "I've been asked if I regretted taking the job since this will be a loss year," he continued. "My response? 'Hell, no.' I'm right where I want to be. I wouldn't trust anybody else." Streeter reminded those assembled of the difference

MFA Stores

Vandalia

Vienna

Walnut Grove

"I have faith in you, I have faith in our customers. I have faith in myself. And I have faith in MFA."

—Bill Streeter

Help welcome Bill Streeter as MFA's new president and CEO.

Attach the official MFA/Target 20 binder clip to your shirt or blouse so that all MFA home office employees will be wearing one during Mr. Streeter's first official week.

If you need additional clips, contact the Mailroom or Communications.

Agriculture is in the midst of uncertainty. Look at the volatility in commodity markets and crop inputs. At MFA, we will be cautious but not overly cautious. We will be active in the marketplace. We will focus on our strengths: increasing sales, maximizing margins, capturing service revenues, and implementing new technology.

There is a tendency in times like these for farmers and industry to retreat. That's not an option at MFA. We will not cut back on product offerings, employee training, or technology adoption. We will continue to develop, implement and innovate as we have for 95 years.

Bill Streeter
Quoted in *CropLife* magazine in 2009

"I've been in wholesale. I've been in retail, I've been in sales. I've been in trouble."

—Bill Streeter

between the breakfast contributions of the chicken and the pig. The chicken is involved, he said. The pig is committed.

"Think of me as the pig," he said. "I'm committed. I've been in wholesale. I've been in retail, I've been in sales. I've been in trouble." That initial relief turned into howls of laughter and waves of applause.

That's Bill Streeter. He embodies leadership. In that manner, he fits the template forged by such MFA leaders as Hirth and Frew. He radiates confidence. He's a can-do individual and people match their performance to his expectations. MFA would not suffer long with Streeter at the helm.

"He was the right man at the right time," said Allen Floyd, now retired, but in 2009 senior vice president of finance and chief financial officer. "Bill Streeter took the bull by the horns," said Floyd. "He brought all the right people together. He outlined all the right steps. There was urgency, not panic. His leadership was unmatched and unquestioned."

Brian Griffith is now senior vice president of corporate operations. He has worked with Bill Streeter for more than 20 years. With a law degree as well as a master's in ag economics, Griffith had been a part of the management team since Bud Frew. He'd seen Streeter operate in good times as well as trying ones. "Under his leadership, MFA

MORSOY SEED
XTRA
R2 44X42N
PS11227
genuity
ROUNDUP READY 2 YIELD
SOYBEANS

not only survived a terrible year, but experienced rejuvenation by increasing sales efforts and providing enhanced services and products to MFA customers," Griffith said.

Making the necessary readjustments would be all the more difficult, because, starting in 2007 and ending in 2008, MFA in conjunction with CoBank had put together a syndicated loan for additional capital. That loan included eight farm credit banks around the country. The instant MFA financial officials recorded a write-down on that fertilizer, MFA would be in default of loan covenants.

"The fact we had all ag lenders, that there were no commercial lenders, made the situation a lot easier," said Floyd. "We immediately made contingency plans." The problem was that the new lenders had no history with MFA. So the banks began tightening covenants enough that they had a say in MFA's response. Streeter bristled under those conditions. While the bankers had a say in MFA's planning, the bankers did not call the shots: Streeter did. "The best way out of difficulty is through it," he announced.

MFA executives outlined a substantial, detailed plan and then worked that plan. "We accomplished what we said we would do," noted Floyd. "We built credibility and trust with those lenders. Long-term that's a good thing."

From Griffith's perspective, none of this happened in a vacuum. Planning, execution and Bill Streeter were at the core. "Some may say MFA got lucky the last couple of years. But, one of Bill Streeter's favorite quotes sums it up well: 'Luck is when preparation meets opportunity.' Bill Streeter is the reason MFA was prepared."

Under Streeter's leadership, MFA emerged from the market collapse wiser, streamlined and more strongly committed to fundamentals. By the end of fiscal year 2010, MFA was back in the black with a profit of $9.7 million, meager but a step in the right direction.

"We've lost money before," Streeter said at the time. "None of us likes it. But, we've also used two decades of profitability to build MFA's balance sheet and strengthen fundamentals by reinvesting in the business. MFA has been a viable presence in the marketplace for 95 years. Make no mistake; we'll be here for another 95."

To achieve that longevity, Streeter focused on assembling and training a professional team at all levels of the organization. With that team in place, MFA's focus would move to sales, margins, revenues and new technology, he said.

As president, Streeter thought big picture. To keep MFA vibrant in the future, the cooperative needed competent people, improved performance, growth and motivated employees ready for the next level. With his sales background, Streeter understood the importance of setting goals. Implementing them would move the cooperative forward. First and foremost, Streeter focused on training and development. Detailed courses needed to be tailored to multiple job levels so that an educated workforce would be waiting in the wings.

To be most effective, he knew, employees must be well trained in the business and well trained in the operating environment. With business metrics mandatory for performance, as many people as possible needed to understand the importance of the balance sheet and its measurements.

Succession planning was a natural benefit of the training program. Too often in years past, MFA had suffered from a top echelon of management close to retirement. Not enough young talent waited on the bench. Bill Streeter was no Fred Heinkel. He wanted talent to tap immediately. "We may not have a depth chart filled out with individual names," said Streeter, "but we are focused on developing a good pool of talent. We want them trained and developed so they'll be ready to step up when needed."

Streeter extended his training philosophy to MFA's corporate board as well. Board trips centered on MFA operating units across the trade territory, as well as site visits at manufacturing facilities, plant foods operations and educational centers like the Danforth Plant Science Center.

He was just getting started.

Finance had a large role to play. "I want to make sure the operating divisions understand finance," Streeter said. "But it doesn't end there. I also want finance to understand the operation divisions and the retail market. Operations and finance have to work hand-in-glove." That understanding across disciplines keeps everyone pulling in the same direction.

Finance covers a lot of ground, Streeter noted. MFA has in place detailed financial and operational plans that roll three years into the future. Not only do the plans cover the company's direction, they also minutely outline marketing and financial objectives. A fan of Dwight Eisenhower, Streeter readily quotes, "Planning is everything; the plan is nothing."

Risk management had been burned into everyone's vocabulary by the 2009 losses. Under Streeter's watchful eye, all operating units in

In 2012 Bill Streeter could announce to the membership that MFA was one year ahead of plan and on sound financial footing. "We are developing targets into 2015 by operating quarter," he said. "These quarterly forecasts are detailed and consist of annual earnings, inventories, investments, acquisitions, debt and payables."

the company developed individual policies that fit into the overall risk-management structure. Risk management at MFA is not confined to obvious areas like plant foods and grain. Also included are components of agronomy, livestock, interest rates and pension funds.

That left growth.

Streeter, who in his previous role in retail pushed for expansion, would also nudge forward and refine the acquisition and expansion process after the delay caused by the initial loss year. MFA's team represented accounting, administration, finance, tax, human resources and operations. With the team in place, MFA expanded its reach in Kansas and Iowa and acquired an important and historic MFA local in El Dorado Springs, Mo. All told, MFA added eight ag retail facilities and a large feed mill and grain facility over the course of 2012 and early 2013.

The next 100 years

By continuing the focus on balance-sheet strength and expansion, Streeter is positioning the cooperative for continued growth, but only if that growth matches strategic direction.

"I've never failed to be awed by this cooperative's market reach and our underlying philosophy of serving our members," he said. "When I look at MFA, I see a streamlined, complex organization devoted to serving customers. For years, we have described MFA as the farmer's vertical integration into the farm supply and grain business. It's the farmer's investment with an opportunity to share in potential earnings. The products, services, facilities and long-range plans are developed with our members in mind."

MFA's devotion to moving products to customers is reflected in the vast array of rolling stock, millions of bushels of grain storage, full lines of supplies, complete livestock focus and a plant foods structure that is second to none.

What's the unifying theme in all this? Simple. Bill Streeter understands the historic reach of MFA. He understands that today's cooperative is built on the strength, vision and effort of thousands of farmers and ranchers—the thousands of individuals who made today possible.

"These pieces are part of a complex distribution and retail system custom designed to be the largest, best maintained and most extensive structure in place in our territory," said Streeter. "And it exists for one reason: To make sure you, our customer, our member, our owner, have access to product when that product is needed. Keep watching MFA. We're poised for great things."

"I've never failed to be awed by this cooperative's market reach and our underlying philosophy of serving our members."

—**Bill Streeter**

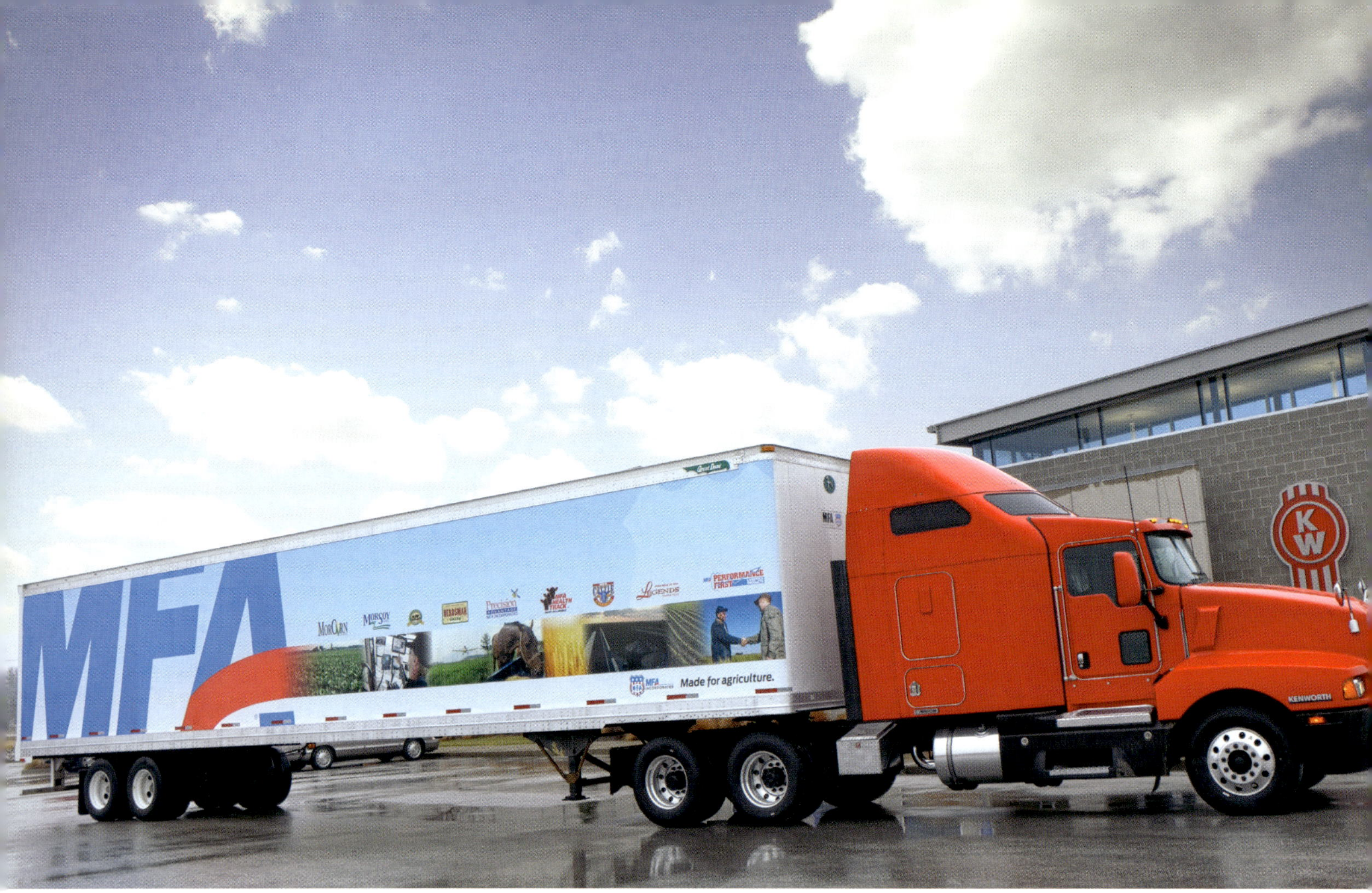
MFA
Made for agriculture.
KENWORTH
KW

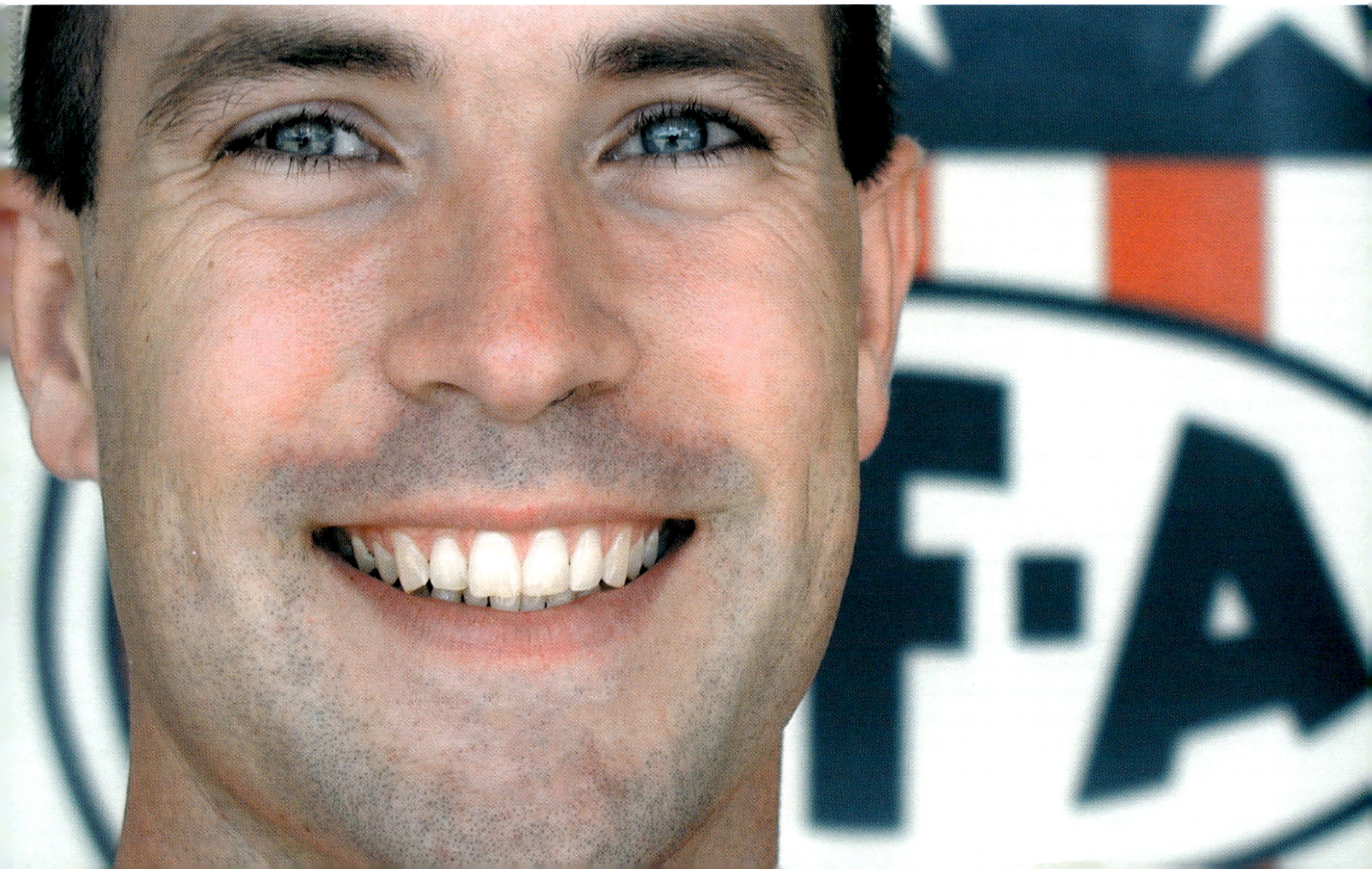

MFA

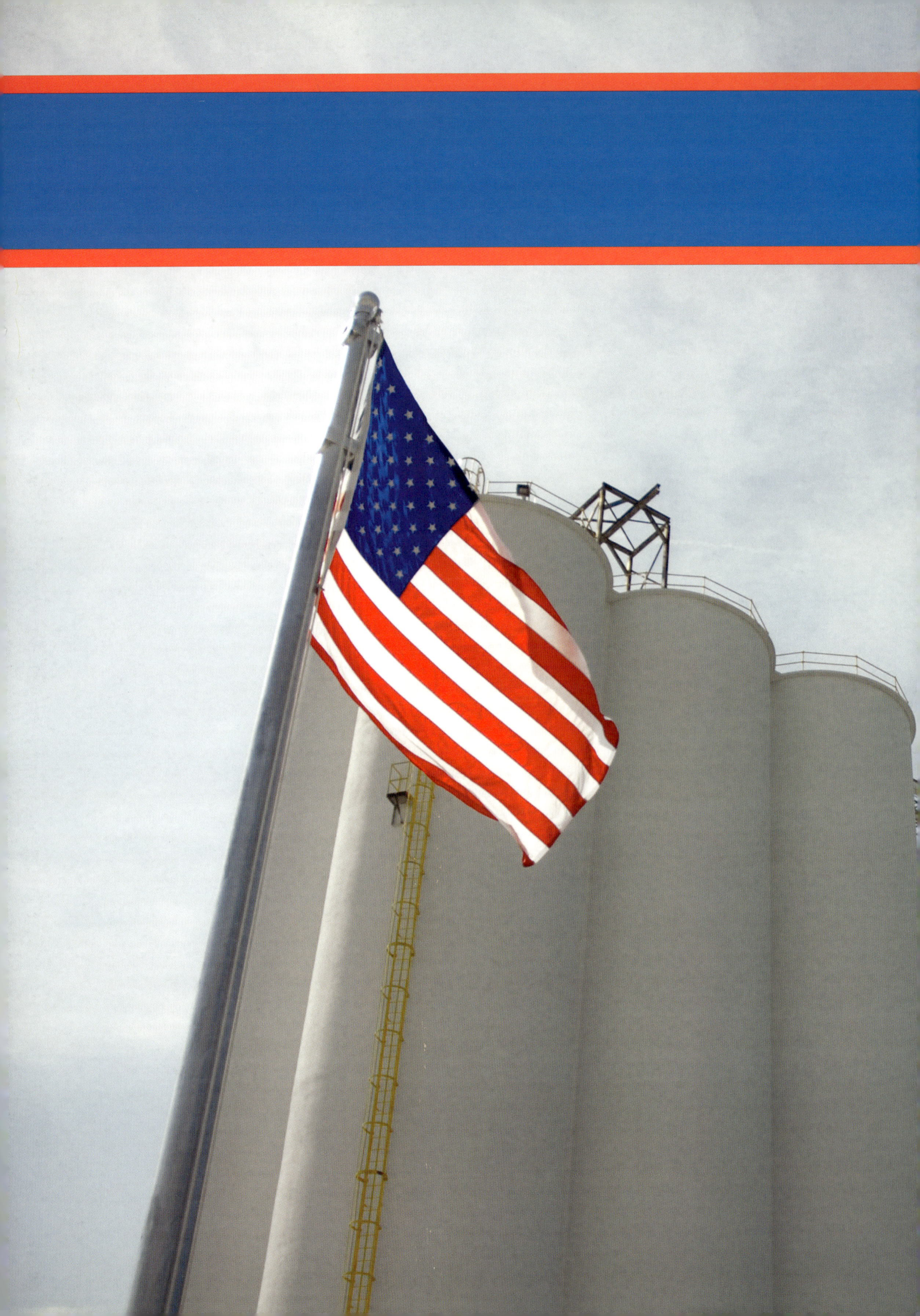

"Keep watching MFA. We're poised for great things."
—Bill Streeter

Photograph by James Fashing

About the Author

Chuck Lay is director of communications for MFA Incorporated and executive editor of Today's Farmer magazine. He joined MFA in 1988. Lay is a graduate of Missouri State University with bachelor's and master's degrees in English and writing. He has twice been named writer of the year for the Cooperative Communicators Association and designated a master writer by the American Agricultural Editors' Association.